Evolutionary Intelligence

Evolutionary Intelligence

How Technology Will Make Us Smarter

W. Russell Neuman

The MIT Press

Cambridge, Massachusetts | London, England

The MIT Press would like to thank the anonymous peer reviewers who provided comments on drafts of this book. The generous work of academic experts is essential for establishing the authority and quality of our publications. We acknowledge with gratitude the contributions of these otherwise uncredited readers.

This book was set in ITC Stone Serif Std and ITC Stone Sans Std by New Best-set Typesetters Ltd. Printed and bound in the United States of America.

Library of Congress Cataloging-in-Publication Data

Names: Neuman, W. Russell, author.
Title: Evolutionary intelligence : how technology will make us smarter / W. Russell Neuman.
Description: Cambridge, Massachusetts : The MIT Press, [2023] | Includes bibliographical references and index.
Identifiers: LCCN 2022052212 | ISBN 9780262048484 (hardcover) | ISBN 9780262376242 (epub) | ISBN 9780262376235 (pdf)
Subjects: LCSH: Technology and civilization. | Human beings—Effect of technological innovations on. | Human–computer interaction. | Computational intelligence.
Classification: LCC CB478 .N435 2023 | DDC 303.48/3—dc23/eng/20230224
LC record available at https://lccn.loc.gov/2022052212

10 9 8 7 6 5 4 3 2 1

Contents

All figures by Tim Foley

Prologue

It appears to be an inauspicious start to your mission. You'll have to make the best of it. You arrive suddenly for your assignment in a lightning flash and a gust of wind. You are crouched down. You are nude. No clothes. No resources. No transportation. No weapons. This is awkward. But you walk confidently toward a nearby biker bar to get what you need. The neon sign reads: The Corral . . . Open. You walk in, and a waitress with a full tray of beer bottles reacts to your state of undress with a mix of surprise and what appears to be a smile of admiration. The bearded bikers with pool cues in hand appear to be somewhat more skeptical.

Some readers will recognize this scenario as the beginning of the popular 1991 action-adventure film *Terminator 2: Judgment Day*. The man on a mission is played by Arnold Schwarzenegger. (He experienced a similar iconic opening scene in the first *Terminator* movie back in 1984.) He is about to get what he needs. What happens next presents some defining imagery for the theme of this book. Schwarzenegger is a muscular man with uncommon sartorial requirements.

Director James Cameron adroitly shoots the scene through Schwarzenegger's eyes. Schwarzenegger is the Terminator, a cyborg time-traveling from the future. When he surveys the environment, his digital visual system overlays relevant data on the scene he is observing. The first screen of data we see as he walks toward the bar with the motorcycles parked out front reads: "Acquire Transport: Priority—1238905D." His visual system is reviewing and reporting on the size, horsepower, and range of each bike. "Scan Mode: Assess Vehicle—Harley Davidson Model 956 Fatboy—Suitable." He enters the bar. "Priority: Acquisition of Suitable Clothing." Finally, Schwarzenegger approaches one of the bikers, played by character actor Robert Winley, who specializes in portraying tough guys and who is about Schwarzenegger's size. The view through the cyborg's eyes reveals the screen data overlay, "Scan Mode: Size Assessment—Initial Match Suitable," with detailed data on neck size, boot size, and so on. Schwarzenegger, in a calm and flat voice, says, "I'm going to need your clothes, your boots, and your motorcycle." There is laughter around the bar. This being Hollywood, Winley mockingly responds: "You forgot to say please," as he blows smoke in Schwarzenegger's face. The screen data reads: "Threat assessment: Scan Carcinogen Vapor." Winley then puts his cigar out on Schwarzenegger's unclothed chest. Schwarzenegger, of course, doesn't flinch. Violence ensues, and in short order the appropriately attired Schwarzenegger, now sporting sunglasses and a few weapons, drives off on the Harley with the grinding guitars of "Bad to the Bone" reverberating on the soundtrack as the movie gets underway.

Not to make too much of this iconic Hollywood narrative fragment, but it reveals several telling points about a critically important next stage in the evolution of smart technologies. The movie depicts a human having a brief conversation with a robotic machine that is humanlike in appearance and in each utterance and movement. The relationship between human and machine lies at the core of science fiction, the core of the present book, and the core of the very definition of machine intelligence. One famous test of machine intelligence, the so-called Turing Test, was developed by computer pioneer Alan Turing in 1950. The test posits that if a human is conducting a conversation with an intelligent agent and is unable to distinguish whether it is a machine or a real human, then the machine has truly reached a level of demonstrable human-level artificial intelligence. The Turing test has dominated thinking about machine intelligence for three generations. I will argue that this highly intuitive model is actually misleading and wrong-headed. But first, back to Hollywood.

What is unique in the movie scene is that the director invites the viewer to experience the exchange from the point of view of the robot and that twist makes it especially important for our central argument. The viewer's first impression is that this data-display thing is interesting. The display provides helpful stuff—the first motorcycle encountered is not appropriate for your needs; the biker dude in front of you is just the right size.

Then you realize that this data-display scenario makes no sense. The central processing system of the robot is digital, of course. Why would one part of the machine type out a text overlay display, in English no less, for the visual system to scan, interpret, and send information back to the central processing system about appropriate resultant behavior? It would be a crazy, convoluted, error-prone, inefficient design. But this is Hollywood, and the task at hand is making an engaging movie. Cue the grinding guitars.

But wait a minute. From a human's point of view, if you had a digital helper, that's how you would do it. Augmented intelligence to help you interpret your environment. A visual overlay of graphics or text. Maybe an electronic voice only you can hear reminds you of critically important information. In my thinking, this surreal movie scene dramatically displays a central element of what I'm going to argue is a critical stage of evolution in the human cognitive system.

Reimagine a scene with two humans communicating. Each person has digital glasses or digital contact lenses with an overlay displaying relevant information. Perhaps the two individuals would be exchanging stories but may have forgotten important details. Or they might be negotiating and need background information about the deal at hand. Or if they were playing poker, each might try to interpret the other's emotional state or might need to calculate the odds of a straight flush given the cards already played. A little extra information would be useful.

We start our journey in this book by recounting a famous scene from an iconic Hollywood action movie. Why that scene? Three reasons, actually. First, the *Terminator* motion picture franchise, reportedly earning about a half-billion dollars in box office revenues, represents a classic and persistent trope in popular fiction's portrayal of the interaction of humans and smart machines—they are at war. Second, as noted, this particular scene offers a rare, perhaps unprecedented audience view from the eyes of the machine rather than the human. It starts to blur the we-are-at-war scenario and psychology. Third, as an innovative narrative invention to distinguish this second movie in the series from the first, Schwarzenegger's character is returning from the future to help the movie's protagonist rather than harm him. Of course, this is an action movie, so we need an antagonist, who is duly supplied in the form of another particularly nasty shape-shifting robot. And as one might predict, the Schwarzenegger character provides help to the human protagonist primarily in the form of brawn rather than brains. But it's a start in thinking in a new direction—a direction that will guide our journey.

This is a serious book about what comes next as human ingenuity conjures up increasingly sophisticated technologies. And there is, as is often the case, a paradox. As we invent new things we are not always sure what to do with them, how to use them effectively, how to avoid having these new gizmos controlling us rather than the other way around. A quick look back at the history of breakthrough technologies reveals that the brilliant inventors, perhaps more often than not, did not fully understand the ramifications of what they were inventing. They were a bit vague about how the technical inventions that would make them famous would ultimately be used. It was true of the printing press, the telephone, broadcast radio, the computer, and the Internet. I would label this the paradox of the brilliant inventor. They imagine their invention as a new thing in an otherwise unchanging world. Big mistake. It turns out these recurring misperceptions provide us some important lessons.

A serious book, yes. But it is great fun to speculate on how human–machine dynamics will evolve in time. There is a gamelike quality to the process of taking current technical, social, and economic trends and trying to extend them into the future. Many such narratives are utopian. Others, perhaps the majority, are dystopian—helpfully warning of misplaced trust in technical systems, unanticipated malevolence, the atrophy of human capacities from technical dependency, or the hubris of the technical expert. Our approach is distinctly utopian in flavor, hopefully not naively so. We will focus on the opportunities that technological advance will afford.

The Purpose of This Book

This book aims to provide you with a new perspective on a cluster of new technologies that lie ahead. This is a bit of a challenge for an author because almost everybody already has a notion of the future and probably a prediction or two of what will become the next big thing. My thesis is that most of the popular next-big-thing candidates are not ill-conceived; rather, they are small parts of a much bigger, more dramatic, and less well understood development. The terms *artificial intelligence* and *machine learning* in particular are fraught with controversy, suspicion, and misunderstanding. These notions project the idea of intelligence and agency as being "out there" in the machine and responsive to the machine's goals rather than our own. Our thesis is that alignment is possible—the machine's goals can certainly be our own. The killer-robot meme is a classic case of humans projecting human traits onto machines. Should we be surprised that when we build robots we instinctively and impulsively make them in the shape of humans?

This book is about a fundamental revolution in the human use of computational intelligence.

It is a revolution in human activity comparable to the invention of language, writing, agriculture, and steam and electric power.

In the complex natural history of human evolution, we survived when many species (most, in fact) became extinct. Humans are not particularly strong or fleet of foot or protected by a natural layer of protective armoring. We survived by our wits, our capacity for social collaboration, our invention of language and physical tools. The human cognitive system hasn't changed significantly in the last 400,000 years. Our brains are no larger. We still have about 100 billion neurons to work with. But now a fundamental shift is occurring as we begin to build tools not just to work with but to think with.

It is *evolutionary intelligence*—this time with selective invention rather than selective survival.

In my view, it is an inevitability. Judge for yourself as we review the issues.

If it is, in fact, an inevitable outcome of current trends, one might rightfully ask, Why is it not obvious, well-recognized, and carefully studied? Good question.

To make my case that it is both inevitable and not commonly recognized, let me draw on a little recent history. Among the last big things in human history that we proudly acknowledge is the Industrial Revolution. Schoolchildren can recite the scenario. Farming and manufacturing were powered by humans and animals and, when geographically convenient, the power of rivers and streams. In the nineteenth century we invented steam power, electric motors, and internal combustion engines. Life was transformed. Cars, cities, suburbs. Modernity. The question we might explore, however, is how well this transformation was understood while it was happening. It turns out that the phrase "Industrial Revolution" was not used until about 100 years after it was well underway. How is that possible? Daniel Bell, writing in the late 1970s, puts it this way:

> It is a rare moment in cultural history when we can self-consciously witness a large-scale social transformation. . . . Few persons realized, when the industrial revolution was beginning, the import of what was taking place. The term itself was coined only a hundred years after the start of the process, by Arnold Toynbee, in 1884, when he gave a set of lectures retrospectively viewing the era that he called "The Industrial Revolution."

We are now not quite 100 years into the digital revolution, and the importance of the computer as socially transforming has been clear and widely discussed from the beginning, even when computers were room-sized monsters programmed by specialists for primarily technical tasks. We should note, however, when speaking of room-sized computers, that IBM, which dominated the early days of commercial computing, did not see the next stage of midsized computers coming and left the field open to new entrepreneurs like Ken Olsen of Digital Equipment Corporation (DEC), who foresaw an alternative to IBM's "big iron." By the late 1980s, DEC had $14 billion in sales and ranked among the most profitable companies in the United States. Ironically, although some of Olsen's engineers experimented with personal minicomputers, Olsen was not impressed and dismissed the project famously noting (in 1977), "There is no reason for any individual to have a computer in his home." So much for self-conscious awareness of major transformations. By the way, like Olson, IBM was late to the minicomputer revolution as well, which created the engineering vacuum to be filled by Bill Gates, Steve Jobs, and others. So this would appear to be the computer industry's version of generals enthusiastically preparing to fight the last war, as the proverb darkly reminds us.

The Structure of This Book

The first chapter lays out the case, identifying the revolutionary capacities these new technologies make possible. The intelligence I identify is not located in a machine or in an individual but rather evolves from the interaction between the two. That's why the Turing test, in my view, is misleading. The phrase "evolutionary intelligence" also identifies these developments as a transformative historical stage, not just a new collection of high-tech machines we should admire. This is not a book about technology. It is about the human use of technology. It about our history. It is about our psychology. It is about our future.

Chapter 2 reviews how the new generation of technology will actually work—the visual, aural, and tactile means of human–machine communication. There is a section on direct-to-brain communication. Electronic communication physically connected to the inner brain is an important issue, but I find that stuff a bit creepy. I am relieved to point out that, unlike the other technologies reviewed, that one is a ways off.

Chapter 3 addresses one of the core issues of the book. I argue that the next generation of artificial intelligence can successfully, even routinely correct for systemic biases in human perception. This compensatory dynamic is what makes these developments potentially revolutionary. I label this "the hard part." Getting humans to take advice, even demonstrably good advice, is not easy.

What could go wrong? A lot. Chapter 4, "Here Be Dragons," is one of the longer chapters. It addresses, among other issues, the questions of privacy, social inequity, and the atrophy of human capacities from dependence on technology.

Chapter 5 takes a step back and gives a historical perspective on evolutionary intelligence. In order to appreciate my argument about major social transformations, it really does help to consider a long-term view of humans and the technologies they invent. My list of evolutionary stages is short. In my version, there are only four truly world-changing revolutions leading up to the current digital revolution. The four stages are (1) the capacity for language, (2) the capacity to settle on land rather than forage, (3) the leverage of machine power, and (4) the belated development of mass education and mass literacy.

When Charles Darwin was wandering around the Galapagos Islands, he noticed some patterns in the local faunas that revolutionized his thinking. These patterns had been around for millennia and were not actually unique to these remote islands. He just needed to take the time to see the evidence before him in a new light. In chapter 6, I make the case that the roots of evolutionary intelligence are all around us today.

Chapter 7 digs a little deeper to explore how evolutionary intelligence may affect different spheres of our lives—through personalized medicine, intelligent romance, digital law, amateur athletics, cyberfinance, and employment.

Finally, chapter 8 returns to the present to address next steps and immediate concerns as the potential success of evolutionary intelligence tomorrow will depend in many ways on what we do today.

So, buckle up. There is a lot to think about.

1
Evolutionary Intelligence

Evolutionary intelligence is the next stage of human evolution as our capacities coevolve with the technologies we create. The wheel made us more mobile. Machine power made us stronger. Telecommunication gave us the capacity to communicate over great distances. Evolutionary intelligence will make us smarter. In my view it is an inevitable process as networked computational intelligence gradually moves from a keyboard and screen on our desk to our technologically enhanced eyes and ears. It may be difficult to imagine how that will work, given our current experiences with technology. So, to metaphorically extend from what is currently familiar, imagine an invisible Siri-like character sitting on our shoulder, witnessing what we witness and from time to time advising us, drawing on her networked collective experience. She doesn't direct—she advises. She provides optimized options based on our explicit preferences. And, given human nature, we may frequently choose to ignore her good advice no matter how graciously suggested. Evolutionary intelligence might be more formally defined as the purposively structured technical mediation of human communication.

The technologies of the Industrial Revolution changed our relationship with our physical environment. Digital technology changed our relationship with our informational environment. Evolutionary intelligence (EI) is a change in our relationship with each other. It is the use of technology to enhance our capacities when we speak or listen. It is not the same as artificial intelligence or machine learning because in those cases agency is relegated to the machine, which then makes decisions and enacts behavior. EI is the enhancement of the human capacity to make decisions. It is augmented and amplified human intelligence. It is evolutionary in the sense that our tools have defined our enhanced capacities for the last 400,000 years. There was no meaningful change in our DNA or cranial capacities as a result of selective survival over that period. Such genetic changes take eons. Our lives have changed dramatically because of our tools.

Note that we have avoided terminology that locates this augmented capacity in humans themselves—some variant of "singularity" or "super smart humans." The intelligence isn't *in* the humans; rather, it is in the enhanced interaction among humans.

Think of it as a routine automated intervention as humans communicate with each other. We can return to our initial example of Arnold Schwarzenegger's action-adventure movie character desperately in need of clothing, transportation, and weapons. A textual overlay on his field of vision advises him every step of the way. Well, yes, Schwarzenegger's Terminator character was not exactly human. But you get the idea.

Sometimes minor technical changes can make a very big difference. Silicon chips didn't do anything new. They did what vacuum tubes did. They did it smaller, cheaper, and more reliably. Modern computers and smartphones, indeed the digital age of modern society, all are unthinkable without silicon chips. Vacuum-tube computers existed but failed often; they required hundreds of square feet, constant maintenance, large amounts of electrical power, and air-conditioning. Silicon chips replaced vacuum tubes. No change in function. Big difference in outcome. As a result, computational intelligence began as a remote room-sized machine, then became a box on our desk, a phone in our hand, and in the near future, it will again transform into a barely perceptible electronic interface between us and our environment.

Skeptical? You are wondering how having Siri whisper in your ear or seeing the projection of some text on your glasses or contact lenses could be much of a big thing. A reasonable question. But in my view your skepticism reflects a narrow, time-bound perspective. Thomas Edison, for example, took such a perspective in inventing his recording device that came to be called a record player. He initially called it a talking machine. Its capacity to record sound was so limited in fidelity that Edison could not imagine that it would be used for recording music. He saw it as a Dictaphone-type device for dictating a letter or a will. He was judging the device by its then-current capacity, rather than its ultimate technical potential. So when I say "a technical intervention in human communication" you are thinking about yourself and your recent experiences with Siri, Alexa, or Cortana. But that produces two errors of perspective. Think instead (1) about yourself a few decades hence and (2) about how the Siris of the world are going to evolve a few decades hence. Both you and Siri are going to change. A lot.

Humans are amazingly elastic. Perhaps the evolved trait that has contributed most to our continued survival and success is our plasticity and adaptability. Once they were available, we were quick to adapt to horses and carriages, later to cars, later to computers and smartphones. These evolved technologies would blend into our daily activities and would simply become a part of how we conduct our daily lives. Horses used to be a daily part of our existence. Some of us still enjoy a good trail ride. But when we need to get somewhere at any reasonable distance, we don't ride a horse. It just doesn't make sense. We drive. In fact, if we're going for a trail ride, chances are we'll drive to the stables and park in the inevitable parking lot.

In the future, when you communicate with a group, with an institution, or simply with another person, it will be technically mediated—it will be yet another exemplar of evolutionary intelligence. And you'll probably give no more thought to the EI than you give to the car ride to the stables. And you'll do that for the same reason you drive. It will have proved to be more effective and efficient. More effective as, for example, language is more effective than gestures and grunts, farming is more effective than berry gathering, and machine power is more effective than animal power. In the pages ahead, we will review how each of these evolutionary developments was in their time utterly revolutionary. And, at the time, each was only vaguely recognized as such. Think of how quickly Zoom-type videoconferencing technologies were adopted during the COVID era—and how quickly they were taken for granted.

Like the first language, first farm, or first machine, early EI will be rudimentary and error-prone. The first language was mixed with gestures. The first farmers still did a fair amount of hunting and gathering. The scattered examples of EI in the next few years may evade your notice.

Imagine in the near future that you are trying to communicate a moral lesson to a frustrated young child who very much wants to possess and not to share. Your initial impulsive reference to universal principles of ethical morality proves to be neither effective nor age appropriate. EI reminds you of an age-appropriate and memorable fable about sharing that turns out to both well received and behaviorally effective. Siri's successors may be thoughtfully drawing on Lawrence Kohlberg's theory of stages of moral development. Perhaps you actually had read some Kohlberg and Piaget in college, but it just hadn't come to mind in the moment. Perhaps in a moment of gratitude you whisper to yourself, "Good work, Siri," or whatever Siri's successor's name is a decade hence.

You are making notes to prepare for a conversation with a colleague. Your job is to persuade the colleague that a proposed project is a good idea. The meeting will be brief. You have developed a list of 22 reasons that the new project is great. You are advised that 22 diverse and diffuse arguments may be counterproductive. In fact, most of the 22 reasons can be summarized in five more manageable, more persuasive, and more memorable points.

For centuries, the protocol has been for the principal dignitary in a receiving line to have an aide at their shoulder, letting them know the names and identities of each approaching guest so that each could be appropriately acknowledged. In gatherings of the future, each of us will have an automated aide who will (if beckoned) provide a similar service. It is as inevitable as the continued existence of receiving lines.

You are at Caroline's Boutique holding the blouse to your chest looking at yourself in the mirror. It looks a little big for you. You are advised that the same blouse with a better fit is available at an establishment just a few blocks away. And it's on sale over there.

You are in a rather heated argument with a friend whose political views you find to be, well, abhorrent. You come up with a killer argument for your side, including persuasive statistics to support your case. You are EI-advised that your interlocutor will not be persuaded in the least by this point and will simply be annoyed. It is suggested that you politely demur. It is unambiguously clear that there is no benefit in frustrating your friend. You recognize this EI advice is, no doubt, correct but choose to ignore it and proceed with your newly developed argument with all supporting statistics, in abundant detail. Your friend, as predicted, is not persuaded. He is annoyed. Somehow you take pleasure in this. You chose to ignore some good advice. This is critically important. EI is advisory rather than determining. Rather than an unavoidable intervention, it is an option, available when chosen. A counselor, not a dictator.

You are typing away and spell "abhorrent" with only one r. Your spell checker corrects it to abhorrent. You hardly notice. You have come to take it for granted. But be careful. Sometimes autocorrect failures, in addition to being embarrassing, are hilarious enough to go viral. Spell checkers and autocomplete algorithms are, of course, current exemplars of evolving, error-prone EI. So is the taken-for-grantedness of it all.

You are at the Department of Motor Vehicles. There is no line. The paperwork is dispatched quickly and accurately. You're outta there, car registration in hand. All the details were electronically exchanged ahead of time, including scheduling your visit to avoid the crowds. You took an extra moment to set up identity authentication so that next time you can conduct business online and avoid the trip altogether.

Uncle Louie whispers a hot stock tip in your ear at a family gathering. Guaranteed inside information. Slam dunk. Can't lose. Gotta act fast. What to do? Turns out Uncle Louie is trying to unload a demonstrable loser. Timely information on the information's source and the source's motivation can be useful.

Evolutionary intelligence is formally defined as the purposively structured technical mediation of human communication. Purposively? What purpose? Now things get particularly interesting.

EI for What Purpose?

Counterpoise. Mediation. Moderation. Intercession. Compensation. This purpose—systematic compensation for the well-known biases in evolved human cognition—is most important. In this analysis, EI offers special promise. The most useful and progressive outcome of these technical capacities is an accrued, refined, time-tested artificial intelligence to advise us when we are systematically misinterpreting the cues in our environment. To inform us of otherwise unanticipated consequences. To correct predictable errors in our psychological estimates of probabilities. (These errors are surprisingly well understood.)

Will various implementations of artificial intelligence be capable of such advisory tasks? Understandably, not at first try, or likely second or third or fourth try, and so on. But in time, certainly.

A future implementation of artificial intelligence that accurately anticipates predictable errors in human judgment? A skeptic might begrudgingly assent. Yeah, maybe. But a skeptic would be quick to point out that the typical human, especially in the heat of action, would predictably ignore any advice given, despite its demonstrable veracity. Is it naive to imagine that a human would not proudly and stubbornly blunder ahead on primordial instincts? This represents an important challenge to our proposed scenario.

So let's draw on a set of present-day case studies. Take the modern motorist . . . that is, while we still have motorists and cars are not yet routinely driving themselves. Video, ultrasonic, or radar sensors flash lights and sound a variety of warning bells and buzzers when objects in our environment come near. Automotive side-view mirrors by their nature have a blind spot. Modern side mirrors monitor the blind spot electronically and flash a warning light on top of the mirror. If the turn signal is turned on in the direction of the side mirror, the warning light flashes more urgently because moving objects in the blind spot would be particularly dangerous in the event of a turn—not a good time to change lanes. Networked GPS software tracks your actual speed, the posted speed, and importantly for many, the presence of law enforcement nearby. GPS systems anticipate traffic jams and suggest alternative routes. When our car advises us what to do, do we do it? Sure, most of the time. In time, we come to understand when signals might be ignored. Our sensors blink and ring like crazy when we park in the garage. We ignore them there. They are doing their thing. It is a small garage. Objects are very close. Currently, the warning distance intervals are preset. In the near future, they will be context dependent and perhaps become a little more laid back in that tiny garage of ours. When Waze or some similar GPS system suggests an alternative route (typically a complex pattern of backstreets), you weigh the alternatives. Is it worth the trouble to save five minutes? Perhaps not. You accommodate. Waze does too. Will some motorists blaze through a speed trap despite the warning? Of course. At their peril. Survival of the fittest. Consider the most primitive and universal of warning devices for vehicular systems—the traffic light. Do drivers pay attention and obey instructions? It turns out that the answer is 99.48 percent of the time, according to one study. Most humans without chemically impaired judgment have come to the conclusion that barreling through red lights is not conducive to survival. There are exceptions. There always are. Some exceptions have life-changing consequences that draw social attention. What goes unnoticed is how routinely humans patiently and thoughtlessly abide a simple red traffic light.

Let's step back for a moment. I am proposing that this revolution is on the same order of magnitude as the invention of language, agriculture, the Industrial Revolution, mass literacy, and computers. That is a daunting measure of significance. How could something like a word-processing spell checker or some sort of real-time electronic debate coach come close to that?

The answer derives from the same historical dynamics that led to the last stages in evolutionary development—selective survival. Tribes that were more successful at coordinated activity and sustaining collective knowledge through language were more likely to thrive and survive. Tribes with agriculture and domesticated animals had better odds than those dependent on hunting and gathering. Better technology wins wars. The selective survival from technological advantage is clear. One is reminded of the mythic drama as brave but doomed Polish mounted cavalry were reported to have engaged German Panzer tanks in the first days of the Second World War in the fall of 1939.

The difficulty with human ingenuity inventing more and more powerful technologies, including nuclear technology, of course, lies in the frailties of human psychology. An angry human with a club can do only so much damage. Angry humans with nuclear bombs, however, represent the ultimate existential threat. Friendly productive communication is one form of human interaction. Economic or military conflict is another. EI represents a potential mediating middle ground between the two extremes.

Science has come to have a relatively sophisticated understanding of human emotion and cognitive biases. Those emotional sensitivities and attentional biases concerning potential threats were critically important to our survival in the grasslands and forests. But they may serve us less well in an era of instant global communication and nuclear armament. We have developed institutions and regulations to help moderate the boom-and-bust cycle that human psychology generates in market economics. We have been less successful in applying similar moderators to international political and economic competition, so it remains an open question. But we are capable of learning from our mistakes. What if we systematically and purposely applied our understanding of our cognitive biases as a corrective—that is, as a compensatory augmented intelligence?

Yes, my argument is indeed that we have to get EI right. Our survival may depend on it.

What are the cognitive biases that EI can compensate for? The list is expansive. Wikipedia lists 200 of them. We'll briefly review one cluster of examples here and return to this difficult problem in more detail in chapter 3.

Philosopher Immanuel Kant liked the naturalistic metaphor: "Out of the crooked timber of humanity, no straight thing was ever made." Political historian Isaiah Berlin liked it, too, and used it to title his well-known collection of essays. Berlin picked up on the crooked timber part but made the case that institutions that protect the prospect of political pluralism could compensate for the frailties of human nature. I'm with Berlin. We can build something straight if we accept and understand the nature of the timber we confront.

Enter Daniel Kahneman. He won the 2002 Nobel Memorial Prize in Economic Sciences for his work with Amos Tversky in developing a series of models of how human cognition works, called prospect theory. Kahneman and Tversky, however, are psychologists, not economists. How did this happen?

The answer is that their psychological experiments on systematic distortions in human judgment have challenged many of assumptions of traditional economic theory, most pointedly that people make rational choices based on their self-interest. Instead, people rely on gut instinct and rules of thumb that focus on perceived fairness, the emotional significance of past investments, and an irrational aversion to loss.

They organized their findings as "prospect theory"—understanding how an uncertain situation, is framed will predictably change how the human cognitive system is hardwired to respond. They demonstrated, for example, that when individuals were asked to hypothetically select the best procedure to cure a deadly disease, most preferred the procedure that "saved" 80 percent of people to one that "killed" 20 percent. They drew attention to the pseudo-certainty effect: the tendency for people to perceive an outcome as certain while it is actually uncertain. Further, they found that in multistage decision making, the uncertainty of the outcome in a previous stage of decisions is often disregarded in subsequent stages. They found that people may be risk-averse or risk-tolerant depending on the amounts involved and how the probabilistic choice is framed.

Why this sudden attention to prospect theory? It is the ideal case study of how to build straight structures with crooked timber. If you can accurately identify the systematic distortions of evolved human judgment, you can compensate for them. Would you like to have a

Nobel Laureate in economics whisper in your ear as you ponder an investment? I would.

EI Is More Than Just Communication over a Technical Channel

Evolutionary intelligence is more than electronic connectivity. Human communication by technical means over great distance was a defining characteristic of the last century and a product of the Industrial Revolution. Marshall McLuhan, among others, speculated that this grid of wires and satellites created a global village and re-created a now global tribal culture. For example, in an oft-quoted 1960 interview for the Canadian Broadcasting Corporation (CBC), McLuhan said:

> These new media of ours . . . have made our world into a single unit. . . . The world is now like a continually sounding tribal drum, where everybody gets the message . . . all the time. A princess gets married in England and boom boom boom go the drums, and we all hear about it; an earthquake in North Africa, a Hollywood star gets drunk . . . away go the drums again. I use the word tribal. . . . It is probably the key word.

Theorists speculated a century before that the telegraph and telephone, through simple connectivity, would have this effect. In *Walden*, Thoreau grumbled the following:

> We are in great haste to construct a magnetic telegraph from Maine to Texas; but Maine and Texas, it may be, have nothing important to communicate. . . . We are eager to tunnel under the Atlantic and bring the old world some weeks nearer to the new; but perchance the first news that will leak through into the broad, flapping American ear will be that Princess Adelaide has the whooping cough.

You might electronically connect a small town in Ohio with a village in North Africa, but would the denizens of each have any way of making sense of that distant community over the horizon? Just connecting is not enough.

EI represents a new kind of connectivity. Empowering connectivity. Connectivity with three empowering and desirable characteristics:

- Intelligent amplification—value-added transmission
- Intelligent translation—reconfigured transmission
- Intelligent transaction—facilitated negotiation

Intelligent Amplification

This is not amplification in the simple sense of making a sound louder. Intelligent amplification is making the message clearer, higher fidelity, with selectively detailed information, with more context and more background.

Communication can be seen as simply the transmission of information. Think of our first efforts at real-time communication over distances—smoke signals, semaphores, Morse-code messages over telegraph lines, primitive telephony and wax cylinder voice recordings. Each technical step provides more detail and a more accurate and complete reproduction of the original message. The transitions are relatively clear cut: high-fidelity sound became stereo, black-and-white television added color and then evolved into high-definition TV (HDTV) and currently ultra-high-definition video.

Traditional television captured the moving image with roughly 500 electronically traced fluorescent scan lines flashed 60 times a second to simulate continuous motion for the human eye. (There were a few more lines and a few less scans per second in Europe.) A large TV set had a diagonal measure just over two feet. These video standards were good enough. Nobody complained that it was too small or insufficiently detailed. We took it for granted. Color was appreciated. HDTV with a million picture elements in each passing frame was sharper and especially appropriate for larger flat panel displays. Ultra-high-definition, promoted as 4K, provides roughly

8 million pixels and approaches the limit of the capacity of the human eye to perceive the difference. Not that folks were complaining that regular HDTV was blurry. Notice that the frame rate is still about 60 frames per second. It has already reached the threshold of human perception. Note that each step is simply increasing the level of detail—the image resolution. That's the first step of this amplification. The next step is adding intelligence.

Intelligent amplification differs from artificial intelligence (AI), which is a conception that makes machines autonomous and independent from humans. Intelligent amplification puts humans in control and leverages computational capacity to augment our capacity for communication and decision making. One could think of the distinction as similar to that between virtual reality (VR) and augmented reality (AR). VR transports the individual to an artificial computer-created environment removed from the real world. AR provides a visual overlay on actual environments to provide relevant information and detail.

Various understandings of intelligent amplification have been proposed and analyzed since the construction of the first computers in the 1940s. One of the first was developed by the English psychologist and a seminal theorist of cybernetics W. Ross Ashby. On the very last page of his *Introduction to Cybernetics*, published in 1956, he writes:

> Now "problem solving" is largely, perhaps entirely, a matter of appropriate selection. Take, for instance, any popular book of problems and puzzles. Almost every one can be reduced to the form: out of a certain set, indicate one element. Thus of all possible numbers of apples that John might have in his sack we are asked to find a certain one; or of all possible pencil lines drawn through a given pattern of dots, a certain one is wanted; or of all possible distributions of letters into a given set of spaces, a certain one is wanted. It is, in fact, difficult to think of a problem, either playful or serious, that does not ultimately require an appropriate selection as necessary and sufficient for its solution . . . If this is so, and as we

> know that power of selection can be amplified, it seems to follow that intellectual power, like physical power, can be amplified.

One can hear the influence of digital pioneer Claude Shannon in these speculative words. Both Ashby and Shannon draw on the notion of successful communication as the reduction of uncertainty. And one can understand that Ashby, like Shannon, was drawn to games and game theory.

Another early proponent of intelligent amplification was Douglas Engelbart (the inventor of the computer mouse) who used the term "augmentation" rather than amplification. His 1962 report *Augmenting Human Intellect: A Conceptual Framework* begins:

> By "augmenting human intellect" we mean increasing the capability of a man to approach a complex problem situation, to gain comprehension to suit his particular needs, and to derive solutions to problems. Increased capability in this respect is taken to mean a mixture of the following: more-rapid comprehension, better comprehension, the possibility of gaining a useful degree of comprehension in a situation that previously was too complex, speedier solutions, better solutions, and the possibility of finding solutions to problems that before seemed insoluble. And by "complex situations" we include the professional problems of diplomats, executives, social scientists, life scientists, physical scientists, attorneys, designers—whether the problem situation exists for twenty minutes or twenty years. We do not speak of isolated clever tricks that help in particular situations. We refer to a way of life in an integrated domain where hunches, cut-and-try, intangibles, and the human "feel for a situation" usefully co-exist with powerful concepts, streamlined terminology and notation, sophisticated methods, and high-powered electronic aids.

Heady stuff. These early pioneers were ahead of their times because when they wrote these observations, computers were still in their infancy and Moore's law had not yet kicked in. Early computational intelligence was capable of checkers rather than chess. But they laid the groundwork for a concept of coevolution, as both computational and human capacities evolve and build off each other.

Most of us have attended meetings in person as well as electronically through an audio-only telephonic conference call or an online video conference. With audio only, one can't help but notice the missing cues of facial expressions, body postures, reactions of others in the room, and perhaps even verbal nuance, which is garbled in low-fidelity telephony, all of which may be important to the process of collective decision making. Even a high-definition video conference call may be insufficient. When you are there in person, you may well direct your gaze at specific individuals at different points to gauge their reactions. You are exercising directed selective attention. You may also be familiar with the participants in the room. You note that one individual seems surprised by a statement. She never seems surprised, in your experience, so this may be a significant sign. Another person looks amused. But you know they always look amused. What gives you the advantage in understanding the meeting's dynamics is your understanding of the context and background. Similarly, if you attend a meeting and you have studied all the participants' resumes in detail, you are better able to interpret their statements or perhaps even their decision not to speak. The details have been amplified.

Could we characterize intelligent amplification as making all the details available to the participant?

Wait a minute. All the details? All the details are too many details. Who wants to make note of every eyebrow twitch during a meeting or the elementary school record of each meeting participant? We just want the relevant information. That's where the intelligence of evolutionary intelligence comes in. It just might be handy to know that two of the other participants at an important negotiation have actually known each other since elementary school. It is difficult at this stage of artificial intelligence to imagine that machines are any good at figuring out which details in a complex flood of fast-moving visual, textual, auditory, historical input are actually important. But machines are good at learning. And if we keep telling them what turned out to be important, they will remember. They will remember better than we do.

Simple amplification is making everything louder or bigger. Intelligent amplification is selective—like an intelligent filter. Amplify the important stuff, filter out the noise. Polarized dark glasses have a simple and mechanical way of minimizing (polarized) glare so that you can better see the environment. Imagine such a visual filter empowered by an accumulated and evolving understanding of what is important in the environment.

One classic case of paying attention to the details in human interaction is the tell. In the game of poker, both amateurs and professionals are likely to bend your ear with tales of how they caught subtle signs of tension or enthusiasm as others at the table read the cards they were dealt. It may be that some players are not conscious of or unable to suppress nervous ticks or sideward glances. It is always possible, indeed likely, that a sophisticated player could intentionally mimic a tell to mislead fellow players. Books on how to read a tell are legion. It appears to be a mini-publishing business in itself. Reading poker faces would seem to be a dramatic example of a human skill rather than something amendable to computational intelligence. But bluffing happens all the time in poker. The game theoretic of intentional misrepresentation and using "data" from each hand to verify whether the tell was accurate might be just the ticket for calculating refined probabilities. In an EI world, presumably everybody at the table will have a statistically accurate estimation of their hand's winning probabilities whispered into their ears. Why not add the benefit of tracking all cards played so far and any physiognomic data available? Perhaps you're thinking that computers will never be good at recognizing facial expressions. That was once said of computers attempting to identify unique faces. However, current facial recognition technology equals or exceeds the capacity of professionals in law enforcement. The key is machine learning. With lots of data and appropriate and accurate feedback, the machine can "learn." (You may have heard of potential biases in facial recognition algorithms. They turn out to be the result of incomplete and thus biasing training data rather than fundamental errors in the software.)

Perhaps you are thinking at this point that EI is going to take all the fun out of playing poker. I doubt it. Quite the contrary. It may move it to a new level. Imagine a player's insistence that their algorithm is better than yours. And there is just plain luck, which can make the difference no matter how well you are advised.

Intelligent Translation

Reliable machine translation from one language into another is extremely challenging. The natural ambiguity of human language mixed with the inexplicable character of many idiomatic expressions contributes to the challenge. Psychologist Gary Marcus notes that most human speech "is ambiguous, often in multiple ways. Our brain is so good at comprehending language that we do not usually notice." John Pierce of Bell Labs was once asked to head up a federally sponsored blue ribbon committee in 1966 to explore the prospects for machine translation. His committee's report was so candid about the challenges given the capacity of 1960s technology that the report set back federal and private funding for translation research for two decades. New technical developments such as neural nets and deep learning currently offer special promise and have stimulated a resurgence of research and application. Most users of the widely used Google Translate and similar neural-net-based systems describe the results as pretty good and getting better, although every user has more than a few stories of hilarious snafus. We have a reliable test of language-to-language translation through back translation. If you move from language A to language B and then get the same result when you translate back to A from the system—mission accomplished. The translation at this point may lack poetic grace and nuance, but it gets the job done.

So that is job one in translation—a highly challenging task—and so far, there's pretty good progress. But we are raising a deeper issue by using the term "intelligent translation." This EI idea is translation that is actually "better" than the original. Neat trick. How could this work?

We have already seen some straightforward early exemplars. Take autocorrect in word processing that retypes obvious typos and suggests grammatical refinements. That is relatively easy to program using alphabetized lookup tables and rule-based analytics of subject-verb agreement and identifying incomplete sentences that lack a verb. In the audio domain, there is auto-tune software that translates the singer's slightly flat A to a right-in-tune 440 cycles per second. These are rule-based processes that only begin to tap the potential intelligence in intelligent interaction. What is the next step?

The next step is context-dependent translation. The answer to "why is the sky blue" should be fruitfully modified when responding to a five-year-old as opposed to a college sophomore in Geoscience 101. Explain to the former that the atmosphere acts like a prism. Explain to the latter that with the phenomenon of Rayleigh scattering for light frequencies well below the resonance frequency of the scattering particle, the amount of scattering is inversely proportional to the fourth power of the wavelength. Well, maybe it would be best to save that for Geoscience 102. But you get the idea.

Context-aware communication is fundamental to successful human interaction. Your answer to the very same question is remarkably different at a family dinner, in a courtroom, or with friends on barstools. Translating your thoughts so that they are effective in a courtroom setting can be challenging. That's one reason we hire lawyers.

Early-stage context-aware translation simply formulates standard information like name and address so that the right text ends up in the correct line on a form. As forms get more complex, like loan applications or college admissions, the requirements of context-awareness are much more intense and, accordingly, of even greater value to the applicant.

We will explore several case studies of context-sensitive and value-added communication in the chapters ahead to describe in more detail where EI is headed.

Intelligent Transaction

Much of human interaction is more than the simple transmission of information from one individual to another. It is dynamic interaction over time, a negotiation, a transaction. Some transactions are pretty straightforward. In Western cultural traditions, in most cases the prices for goods and services are fixed and posted, and potential buyers simply decide whether or not to take the offer. In the Middle East and parts of Asia, custom requires much more of a back-and-forth discussion of value and price, perhaps over a cup of tea. Western traditions include complex patterns of intermittent sales and selectively applied discounts, and for expensive transactions such as cars and homes, a more complex back-and-forth negotiation is common.

We are all aware of how digital systems have become more and more important to the world of transactions—electronic investing, banking, credit cards, and e-commerce, among other examples. We take it for granted. Roughly a third of transactions in the United States are made with cash (typically purchases under $25) and the rest with credit or debit cards or mobile devices. It varies dramatically by country. Mexico uses cash over 80 percent of the time, South Korea only 14 percent of the time. But the trend toward electronically mediated transactions is clear. The security of these transactions is stable with fraudulent transaction percentages surprisingly low—well under one percent. It seems to be well managed. The risk is generally covered by the credit card companies as they see it to be in their interest. As I say, electronic transactions have long been taken for granted and treated as business as usual.

But these are, for the most part, simple fixed-price transactions with willing buyers and sellers currently in a well-organized market with both buyer and seller well established in the payment system. The corresponding administration of goods and services after sale is managed by independent systems of fulfillment and delivery. So the transaction is electronic and secure but not exactly "intelligent." Where does that come in?

Intelligent transactions take advantage of expert knowledge about the products and services in play, the competitive offerings, the relevant histories of the current market players, the contractual basis of the transaction, and the best information on the future of the marketplace.

Look at it this way. If you are in the market for a pack of chewing gum, you don't need an expert consultant in confectionery products and bubble gum futures markets. If you are in the market for new job, or stocks, insurance, vehicles, real estate, vacation packages—that is, more complex, larger-scale endeavors, you just might take advantage of some specialized expertise. It's worth your while to explore the options and learn a little more about the alternatives. As EI makes mediated networked intelligence more immediate, inexpensive, and readily available, the use of intelligent transactions is no longer limited to the big-ticket items. It gradually migrates down the continuum from big ticket to medium ticket to little ticket. For a new home, we are likely to consult real estate agents, mortgage bankers, insurance brokers, building inspectors, and of course, real estate lawyers concerning contractual details. For a car, we check some websites and ask our next-door neighbor Victor who knows a lot about cars. For a washer-dryer, maybe we consult Consumer Reports, maybe not. You may not want to take the time to calculate relative machine cost against energy cost savings over time (given your usage and your local energy costs, with appropriate discounts for inflationary trends.) But if such information is reliably and readily available, why not?

At the beginning of this chapter we reviewed a little scenario of shopping for clothes at Caroline's Boutique. Fashion choices can be pretty subjective. They can also be pretty expensive. It is also a domain that may engage a great deal of expertise. Some well-to-do consumers hire experts as personal shoppers and stylists. In the future, that expertise may not be quite so limited to the well-to-do.

We also posited an enthusiastic but likely ill-advised "hot stock tip" from Uncle Louie. In the future, Uncle Louie would expect you to routinely (and electronically) run his recommendation against the experts. Who is enthusiastic? Who is not? And why? How does his recommendation stack up against your tolerance for risk? EI does not replace decision making. It enhances it. Your car's intelligent networked navigation system doesn't tell you where to go. You decide where to go. Once you do, it can help you get there efficiently.

Inventing the Future

As you read this, there are perhaps several hundred creative scientists, inventors, philanthropists, entrepreneurs, investors, and science fiction authors trying to figure out how this especially challenging implementation of artificial intelligence might work. In another five or ten years, there will be many thousands hard at work as the ultimate real-world contours of evolutionary intelligence begin to take shape. Currently, the diverse efforts remind one of the blind-men-and-elephant adage as a variety of terminologies and analytic models compete for attention, each emphasizing different pieces of the puzzle.

At the MIT Media Lab, for example, Pattie Maes and Rosalind Picard have been pushing the boundaries that have traditionally separated human and machine. Maes was one of the first to explore in practical terms how an intelligent agent could augment human capacities. Well before the release of Google Glass, her students wore video cameras that provide computer-aided feedback on alternative products as they shopped. She championed the terminology of human–computer symbiosis, assistive augmentation, and technology as our "sixth sense." She encourages us to get comfortable with what she sees as our certain future as natural-born cyborgs, employing our innate human capacity to incorporate new tools into our existence. She points to our increasing dependence on our cell phones as the one of the first steps in this inevitable progression. Her TED talks continue to draw international attention. Rosalind Picard's preferred terminology is affective computing. Her group at the lab has demonstrated that computers can accurately assess the emotional component of human input, particularly facial expressions and vocal inflections as well as gesture and posture. Both Maes and Picard have spun off successful ventures to commercialize their ideas.

Perhaps the most prominent of the public intellectuals who have been exploring this terrain is inventor and author Ray Kurzweil. He has popularized the term *singularity* to characterize the rapid increase in artificial intelligence transcending the limitations of our biological bodies and brains. At the point of singularity, which he estimates to be about 2045, there will be no distinction between human and machine. In his vision,

> information-based technologies will encompass all human knowledge and proficiency, ultimately including the pattern-recognition powers, problem-solving skills, and emotional and moral intelligence of the human brain itself. . . . The Singularity will allow us to transcend these limitations of our biological bodies and brains. We will gain power over our fates. Our mortality will be in our own hands. We will be able to live as long as we want. . . . We will fully understand human thinking and will vastly extend and expand its reach. By the end of this century, the non-biological portion of our intelligence will be trillions of trillions of times more powerful than unaided human intelligence.

A number of other futurists share his vision but not his enthusiasm. Oxford philosopher Nick Bostrom, for example, draws our attention to what he ominously calls the likely "treacherous turn of superintelligence." Artificial intelligence behaves cooperatively when it is weak and learning its way around. But when the AI gets sufficiently strong, he insists, it will strike without warning or provocation to form a superintelligent singularity to optimize the world according to its own criteria. Bostrom's book *Superintelligence: Paths, Dangers, Strategies* was a New York Times bestseller in 2014. It was followed by MIT physicist Max Tegmark's *Life 3.0: Being Human in the Age of Artificial Intelligence* in 2017. Tegmark uses the terms artificial general intelligence (AGI) and "information explosion," a phrase from early computer theory. Berkeley computer scientist Stuart Russell published *Human Compatible: Artificial Intelligence and the Problem of Control* in 2019. Russell believes we need to steer AI in a radically new direction if we want to retain control over increasingly intelligent machines and proposes a set of algorithms he calls "inverse reinforcement learning" to maintain control. We will review their work in chapter 4. These visionaries often attend each other's conferences and workshops to compare notes. Inevitably, they pool the various estimates on when machine-based superintelligence will become a reality. At a meeting in 2015, their pooled estimate was the year 2055. After they met again two years later, the estimate was revised to 2047.

2

Communication Like Never Before—How Would EI Actually Work?

In the prologue, we started out with a lightning flash and a gust of wind as Arnold Schwarzenegger demonstrated a visually oriented version of evolutionary intelligence (EI) at the Corral Bar in the first frames of the motion picture *Terminator 2*. We noted that although the internal electronic communication within a digital cyborg would never be designed like that, such a visually oriented system would be ideal for providing real-time information to actual humans.

EI responds to the all-too-frequent situation reflective of the human condition—an individual in dire need of information.

There is the actor who has forgotten her lines, so a prompter whispers the missing words from offstage. Picture a struggling student reading test answers that have been written on the palms of his hands, or a nervous job candidate at the interview with what-to-say cues on his shirt cuffs. Or, returning to Hollywood, there is a classic exchange in the 1987 movie *Broadcast News* as the brilliant news producer (played by Holly Hunter) tells the handsome but slow-witted news anchor (played by William Hurt) how to interpret some breaking news while he is on air by speaking to him through an earpiece.

Hurt: You're an *amazing* woman. What a *feeling* having you inside my head!

Hunter: [smiling] Yeah—it was—an unusual place to be.

Hurt: It's like—indescribable—you knew *just* when to feed me the next line, you knew the m—second before I needed it. There was, like, a *rhythm* we got into—it was like—*great sex*!

What it was was great communication.

Imagine the gambler trying to beat the odds at roulette being advised by a spin-counting compatriot through a secret earpiece. No need to dream that up. It really happened. It was the summer of 1961. MIT mathematics professor and notorious gambling enthusiast Edward O. Thorp, author of a best-selling book on blackjack odds called *Beat the Dealer*, talked fellow professor and world-famous information theorist Claude Shannon into a real-world experiment in mathematical prediction. These guys created a minicomputer the size of a cigarette pack and a digital controller built into a shoe and practiced with a roulette wheel in Shannon's basement for months. They analyzed the speed and timing of the roulette ball, and the world's first wearable computer could predict with reasonable odds (they calculated a 44 percent edge) which of eight segments of the wheel the ball would end up in. So, in August 1961, they set off for Las Vegas. Thorp's wife Vivian and Shannon's wife Betty stood as lookouts. Betting was a two-man operation. Shannon observed the ball passing a reference point and manipulated the computer with his foot. Each man wore an earpiece that received tones that signaled which of the eight octants of the wheel was predicted. Thorp did the betting while pretending to be distracted by his notebook. Their bets were relatively small, but the chips added up to quite a pile. The casino never caught on. The professors and their wives left with a modest profit. They were not there to bring down the house. They primarily wanted to make a point. In a few years, Nevada would ban the use of devices designed to predict the outcome of games or aid in card counting.

EI in 1961 turned out to be quite a challenge. They used thin stainless steel wires painted a flesh color to connect to the earpieces. The wires ran down the neck and through their clothing to the radio receiver. The wires weren't obvious, but they were delicate and tended to break, which Thorp called "the Achilles' heel of the system." "So," Thorp says, "we'd bet for a while and then a wire would break, and we had to go back to the room and take the person who was doing the betting, namely me, apart and solder things together and hook me back up." In his 1998 paper "The Invention of the First Wearable Computer," Thorp couldn't resist noting that "once a lady next to me looked over in horror. I left the table quickly and discovered the speaker peering from my ear canal like an alien insect." If you are interested you can examine the original contraption, earpiece and all, at the Computer Museum in Boston.

We have come a long way from 1961. In this chapter, we review the probable physical techniques for routine real-time EI in the future—basically a diverse variety of wearable computers. In the medium and far-term future, we can expect direct-to-brain communication taking various forms. We take a brief look at that as well. For the near term, we have the big three: (1) visual, (2) auditory and (3) tactile means for communicating and computing.

Visual

Several versions of visual EI are already here. They have come to be called augmented reality—AR for short. Whip out your smartphone. Turn on the camera. Any number of apps are available to overlay text and graphics on the reality you view through the camera's video. With the help of your phone's GPS and high-speed connectivity, you can download information about your environment. The history of the building in front of you. Directions to the nearest Starbucks. What kind of an oak tree is that? Pattern recognition is easy. It's a white oak. Did you know that oaks produce more than 2000 acorns every year, but only one in 10,000 acorns will manage to develop into oak tree? Well, now you know. Your reality has been augmented.

We take visual augmentation for granted in sports television. If you're attending a live football game, all of a sudden you may notice that you really miss the computer-generated first-down line you expect with TV. For live games, you may need to rediscover the chain-crew guys holding the posts marking the down on the sidelines. Or at a live baseball game, you need to imagine the strike zone outline overlay sometimes used in televised baseball coverage.

You see someone who looks familiar. You take out your phone and point it at them. You take a picture and run the facial recognition app. It's Marvin who had the cubicle next to you several jobs ago. Marvin recognizes you and wonders what the hell you are doing. It's awkward. That awkwardness turns out to be key to the Google Glass story, one uneasy step backward as we move forward toward EI.

Google Glass has many fathers, including Google founder Sergey Brin, Google X chief Sebastian Thrun, "Captain of Moonshots" Astro Teller, and former Apple designer Tony Fadell. But if the glass project has a patron saint it would have to be University of Toronto professor and MIT Media Lab graduate Steve Mann. Mann has a thing for video cameras. He has pretty much had a video camera attached to his head for the last 40 years. He has his own current version of smart glasses he calls EyeTap. But back in the early days, he had to lug around 80 pounds of primitive video equipment. From 1994 to 1996, Mann continuously transmitted everything he experienced to his website in real time for others to view. (There was no Wi-Fi digital data network then, so Mann put an antenna on the tallest building at MIT to create his own network.) As I say, this guy's got a thing. At MIT you need to put in some extra effort to be viewed as a nerdy guy. Mann somehow managed. It was part of a personal philosophy to use technology to enhance the capacity of the individual to observe his environment. He conducted numerous experiments in making the sound and radio waves in his environment visible. He detests artificial insertions into reality such as Pokémon Go. "I want to see truth, not fiction" he insists. But the environment outside MIT has not always been so solicitous. While ordering at a McDonald's in Paris, he was attacked and thrown out by the staff who were upset and offended by his headgear, which they felt violated their sense of privacy. This sense of violation turns out to be a key issue for wearable technologies in general and Google Glass in particular.

It seemed obvious to many. As computing power increases and gets smaller and smaller, computation moves from the desktop to the laptop to the smartphone and smart watch and . . . to your eyeglasses. Why not?

Around 2010, Google cofounder Sergey Brin started spending most of his time in the newly formed and secretive new-project skunk works called Google X. The idea is to invent and launch "moonshot" technologies that aim to make the world a radically better place. They define a moonshot as the intersection of a big problem, a radical solution, and breakthrough technology. One of the very first moonshots was the development of smart glasses. The early prototypes weighed eight pounds, but the initial commercial version released in 2013 weighed less than an average pair of sunglasses. As news of the secret project leaked out, Brin decided to release the current version commercially, although the display had rather limited resolution and the battery lasted only a few hours. Early users (dubbed explorers) would provide feedback to improve the concept, he posited. It turned out to be a mistake. Glasses struck many as more dorky than cool. And most important, those around the Google Glass wearer felt they were being video-recorded against their will. A big problem.

Take, for example, a *New York Post* piece from 2014:

> But not everyone is going Glass. Critics argue that the flashy gizmo is both pretentious and intrusive, letting wearers take photos with a simple wink of the eye. "I don't see why anyone feels the need to wear them," says 30-year-old Pete, who works in financial research and declined to give his last name for professional reasons. He recently spotted a man with Glass on the subway. "Was he reading his emails, watching an old episode of 'Game of Thrones' or recording everyone? Just reach into your pocket and get your phone!" "Glasshole" has become the term du jour, and outrage has spiraled so out of control that San Francisco has seen a series of reported attacks on users . . . Google provides its own manual on how not to be a Glasshole—including "Don't be creepy or rude"—social navigation can be tricky.

Non-wearers suspected the damn things were recording all the time. They're not. There is a prism reflection on the lens to indicate a video is being taken.

So why the widely accepted conclusion that Google Glass was a big fail? In part, it is a mismatch between Google-sized expectations (and the attendant hype) and a more modest test model. They were a beta release. The release was clearly a tactical mistake. Priced at $1,500 with low-resolution imaging and a two-hour battery life, the privacy miscues were only part of the problem. The display and communication functions would have worked fine without the camera. And the Jetson-style headgear could have taken the form factor of normal glasses. The real issue was captured in a TED-style talk by smart glass pioneer Mark Spitzer. His talk was mostly about technical issues, but there was a throwaway line tucked in the middle of his presentation. It turned out to be a critical insight. He said simply: "There is no killer app for smart glasses." No primary function. No real value added. Mostly a novelty.

That insight helps to explain the 2019 rollout of Google Glass II called Google Glass Enterprise Edition. The idea (perhaps not quite a killer app) is to enhance the worker with a hands-free set of visual instructions for various physical work processes like manufacturing assembly, physical navigation, and even surgery. But I remained puzzled. Processes like manufacturing assembly become routine very quickly. And I don't know about you, but I would prefer that my surgeons know their way around without needing external guidance.

Perhaps a simplistic notion of a killer app is part of the problem. It presumes a particular and most often narrowly defined function. Our theme in this book is that EI itself functions as a killer app. But it is not a singular activity. It is a more general capacity for enhanced and empowered communication. EI is useful whenever an interaction can be better informed for all parties involved. Important things like the best price available for a major purchase. Less critical things like the lyrics to a song.

One lesson from the Google Glass adventure is that headgear contraptions with video cameras strike most of us as a bit creepy. The privacy issue is central and complex. For the moment, we will set privacy issues aside to be addressed in more depth in a later chapter and simply review the physical properties of visual EI.

Imagine a humdrum pair of glasses. Nobody notices them. The wearer and others take them for granted. Their function is clarifying vision. Reasonable enough. Smart glasses, however, have the capacity to project a visual overlay image of graphics, text, even video visible to the wearer but otherwise unobtrusive for others. Although one could design smart glasses without video input, it would be an obvious help in interpreting the environment. One could imagine asking the smart glasses to tell you what kind of car that is. Or what the shortest route to 20 Ames Street is from here. The asking could be an auditory process, much like our present-day routine interaction with Alexa, Siri, or Cortana. It could be preprogrammed so that when I get to the door, it reminds me of the access code. Or preprogrammed to give me someone's name and job title when facial recognition kicks in. Or you might tap the glasses and have the artificial intelligence of the system anticipate what information you might be seeking.

I know. Even with the privacy issue set aside, it all seems a bit creepy, mechanical, and artificial. True, but prescription eyeglasses are mechanical and artificial, yet we take them for granted as a natural aid to human visual perception. For the last few years, our cars have been beeping and flashing at us when an object or wall is near when we park. Automotive proximity sensors represent an intelligent and artificial aid to our capacity to negotiate our environment without bumping into things. They don't steer the car; rather, they provide information to the driver. Seems pretty reasonable.

When voice communication with the head-mounted display is not appropriate, there are keyboards and single-hand keyed input devices. Our friend Steve Mann developed multiple keyed devices and gloves that permitted tactile input while on the go decades ago. Modern commercial versions have much lighter, flexible, and comfortable form factors.

You're playing chess with a friend. A friend who wears glasses. Fair is fair. You want to compete with your friend, not your friend's computer. Is there a problem? Not necessarily. We have all sorts of competitions that engage our technologies as well as our personal skills. Take auto racing, for example. There is NASCAR and Formula 1 racing. Each have their own rules on what is permitted, and drivers and mechanics do their best to optimize the result, given the rules and the limits of the technologies. It's competition among human–machine competitors. And one of the latest crazes is stadium video game tournaments. Human competition in technically created virtual environments with immense crowds cheering the world-class gamers on. Set up the rules. Play by the rules. May the best avatar win. Just agree with your chess partner whether or not Big Blue will be giving anybody some advice.

The next stage after smart glasses? Wait for it. Perhaps it is obvious. The next stage after smart glasses is smart contacts. There are numerous challenges in getting elements sufficiently miniaturized, powered, and transparent. Sony, Google, and Samsung have filed patents related to smart contact lenses. The Google project begun in 2014 was focusing on measuring glucose levels in tears, which moved them in a somewhat different direction and ran into technical difficulties. Babak Parviz and Harvey Ho at the University of Washington were hard at work on bionic contact lenses a decade ago. Both researchers got distracted by other challenges (Parviz worked on Google Glass and is now at Amazon), so the research they initiated may have gone underground for a while as patent issues are worked out. They developed a one-pixel display and a working Wi-Fi antenna to communicate to supporting wearable technology such as a watch or smartphone. Primitive but promising. One version of the bionic contact lens attempted to use the blinking reflex to recharge the battery. There are some very talented folks working this through, but at the moment they are keeping very much to themselves.

One of the first smart contact lens research teams to go public is Mojo Vision. This start-up based in Saratoga, California, uses virtual reality headsets to demo the lens it developed because approval by the Food and Drug Administration (FDA) is still a few years out. Mark Sullivan, a journalist at *Fast Company*, described the demo as follows:

> When I looked into the user interface of Mojo Vision's augmented reality contact lenses, I didn't see anything at first except the real world in front of me. Only when I peeked over toward the periphery did a small yellow weather icon appear. When I examined it more closely, I could see the local temperature, the current weather, and some forecast information. I looked over to the 9 o'clock position and saw a traffic icon that gave way to a frontal graphic showing potential driving routes on a simple map. At 12 o'clock, I found my calendar and to-do information. At the bottom of my view was a simple music controller.

He reported on the actual lens prototype this way:

> [It] focuses its light on a tiny indented area of the retina at the back of the eye called the fovea, which we use to detect the fine details of

> objects right in front of us. This little indention takes up only about 4% to 5% of the area of the retina, but it contains the vast majority of its nerve endings. It's thick with photoreceptors that convert light into electrochemical signals, which are then transferred through the optical nerve to various vision centers in the brain. Mojo lens will contain a supporting cast of microcomponents. The first versions will include a tiny single-core ARM-based processor and an image sensor. Later versions will add an eye-tracking sensor and a communications chip. At first the lenses will be powered by a tiny thin-film, solid-state battery within the lens . . . the battery is meant to last all day and will charge in a small case that's something like an AirPods case. Eventually, the lenses might get their power wirelessly from a thin device that hangs loosely around the neck like a necklace.

Perhaps you are wondering if my thinking is old school and out of touch. All this discussion of smart glasses and bionic contact lenses is Model T image-making. You know that because you've been to the movies. Perhaps visual EI will not require glasses or contacts. The graphics and text will hang in the air like holograms. You recall that in the original *Star Wars,* R2D2 projected Princess Leia's image into the center of the room as a flickering video with the message: "This is our most desperate hour. Help me Obi Wan Kenobi, you're my only hope." Or perhaps you were thinking of Tom Cruise in *Minority Report* moving images of crimes scenes and mug shots through the air with his gestures and high-tech gloves. Hollywood loves the idea of images floating in the air, and the illusion is easy to achieve in film and video with computer graphics. But there is a problem with that sort of thing in the real world. Hanging in the air is not how the physics of photons work. Photons don't shoot from a source and then decide to turn around in midair and head toward your eyes. They need to bounce off a surface. Magicians in live performance since the nineteenth century have used variations of the Pepper's Ghost trick. The performer uses a glass or plastic surface to impose a reflective ghostly image "hanging in the air" when carefully illuminated from the side of the stage. The illusion requires a barely perceptible surface to reflect the photons.

What about a hologram, you ask. A holographic image uses a recording medium with a prerecorded image to generate the sensation of a three-dimensional scene to the naked eye. In its pure form, holography requires the use of laser light for illuminating the subject and for viewing the finished hologram. There exists a lower-quality little sister technology called the rainbow hologram that uses natural light for viewing and has found use in security applications for credit and identity cards. In both cases, you have to view the holographic plate to see the image. Again, no light beams actually hanging in the air. So until the physics of photons changes, or we get better at creating force fields, we will require some physical semitransparent surface between the viewer and the environment—and for the immediate future that would be physical screens, eyeglasses/headgear of various sorts, and contact lenses.

Auditory

Perhaps the most common early reflection of EI in our day can be found in the imposing triumvirate of corporate voices—Siri, Alexa, and Cortana—each available in the competitive smart speaker environments of, respectively, Apple, Amazon, and Microsoft. (One has to wonder what they would say to each other if placed in the same room.) Like the EI trajectory itself, smart speakers were inevitable once the technologies of ubiquitous wireless connectivity, reliable speech recognition, and acceptable artificial speech all converged. Want some intelligent help in a wired-up home? Just ask. That is, ask a question after cueing the activation word, typically the smart assistant's name. We have already some real-world evidence of what people want to know—that is, how the use of EI may play out. How do folks in a workaday world use the miracle of computational sophistication to enhance their relationship with and understanding of their environment? It's sobering.

- Set a timer: 85 percent
- Play a song: 82 percent
- Read the news: 66 percent
- Set an alarm: 64 percent
- Check the time: 62 percent
- Tell a joke: 60 percent
- Control smart lights: 46 percent
- Add item to shopping list: 45 percent
- Connect to paid music service: 41 percent
- Provide the traffic: 37 percent
- Add an item to your to-do list: 33 percent
- Buy something on Amazon Prime: 32 percent
- Control smart thermostat: 30 percent
- Play children's music: 29 percent
- Check or add an item to calendar: 21 percent
- Other: 20 percent
- Spell something: 18 percent
- Call an Uber: 6 percent
- Connect to phone by Bluetooth: 4 percent

This is from a 2016 Experian survey of Amazon Echo users. One has to admit it is strikingly realistic. Most people don't have "discover the meaning of life" or "explain Heidegger's classic philosophical tome *Being and Time*" as top of mind. Practical patterns of use represents an important question, and we will address the nonuse, unthinking use and misuse of EI in chapter 4 ("Here Be Dragons").

For the moment, let's take a look at the physical practicality of audio-based intelligence in day-to-day environments. The auditory equivalent of the eyeglasses, projectors, and screens are transducers—technologies that convert electrical waves to audio waves. Examples are speakers, headphones/earpieces, and bone conduction. Speakers provide sound to the environment for all to hear; headphones/earpieces and bone conduction provide for private aural communication to the individual listener.

Many of us routinely walk, drive, exercise, and even work with earbuds bursting with music, news, audiobooks, or podcasts. Earbuds are often wirelessly connected to a smartphone by Bluetooth technology, so the music is automatically paused when a phone call comes in. At that point, the headset activates the microphone and facilitates the two-way conversation. If so inclined, the user can use auditory commands to pause the music or speech, change sources, or stop the audio stream altogether. Let's call it pre-K EI. The earplugs or earphones are intrusive. The audio stream just plays what it is told to play. No adaptive intelligence to the environment or to the needs or aspirations of the user. No awareness of the environment. There's limited interactivity in that it can pause and rewind and can interrupt for a phone call, but that's about it.

The future of mobile audio communication will be unobtrusive—perhaps bone conduction audio from normal looking glasses. Perhaps something like a modern micro hearing aid. The level of interactivity and activity will be set by the user. Perhaps the user would welcome suggestions on how to get to Ames Street more directly. Perhaps not. Perhaps the user would like advice on alternatives to a proposed purchase. Perhaps not. Most often, EI awaits until bidden.

Tactile

We are working our way through the three primary senses—sight, hearing, and touch. We have tactile communication from our cell phones now when we set them on vibrate. The tactile sensory path is much more limited in bandwidth. Different frequencies and different patterns of dots and dashes can be relatively easily perceived by the user. But in all likelihood, tactile will most likely and most often provide an alert function rather than a full-scale communication channel.

The tactile channel is used in braille. Extensive training is required to correctly interpret the patterns of six tiny electronically raised or lowered rods to simulate printed embossed braille, so its use will likely be limited to special circumstances. There is a fair amount of futuristic research on human–machine communication using tactile channels for those of us with impaired vision. One could imagine that some of this work will result in general-purpose interfaces as well. Researcher Susan Lederman, for example, reports that our fingers can reliably distinguish up to 40 distinct areal tactile display patterns (lines, dots, stars, boxes, etc.) from a portable display panel or perhaps ultimately electronic gloves. In all likelihood, these somewhat more exotic interfaces will supplement rather than replace sound and vision if those senses are not impaired.

Direct Neurological Communication

Recall once again the lightning flash and gust of wind as Arnold Schwarzenegger manages a rather dramatic entrance in *Terminator 2*. We drew attention to the words projected as a textual overlay on his field of vision, guiding him on his mission. The scene is visually engaging and very Hollywood. But, we also noted, it is a crazy way to communicate. Why type out English sentences on a screen that would need to be scanned and interpreted if it is otherwise possible to communicate directly and digitally with the cyborg's central processing unit? Real human beings also have a central processing unit—the amazing organ we call the brain. Is direct communication to the brain possible, sidestepping the traditional sensual inputs of sight, sound, and touch we have been reviewing?

The answer is yes. The field of direct-to-brain communication research has been relatively active for the last several decades and is typically termed the brain–computer interface (BCI). Direct communication to the brain is indeed the stuff of awe-inspiring science fiction, but here and now in the real world, it remains at an extremely primitive stage of development. Not much awe yet.

There are a number of reasons. Primary among them is that we really don't understand how the brain works. As the saying goes, it's complicated. Cosmologist Carl Sagan famously remarked that our brains are even more complicated than our galaxy. We estimate there are about 87 billion neurons in a typical human brain connected to each other, with axons and multiple dendrites generating about a thousand trillion potential connections. How do we make sense of a number that big? One observer pointed out that is equivalent to the number of seconds in 30 million years. Connections are important because it is not simply the individual neurons but the connections among them that represent our memories and thoughts.

Current technology provides for two primary noninvasive means to derive a few clues about what is going on inside the brain: functional magnetic resonance imaging (fMRI), which uses a massively powerful magnetic field to assess which areas of the brain are most active, and electroencephalography (EEG), which places sensors around the skull (in what looks like an electrified bathing cap) to assess electrical waves generated by brain processes. fMRI has the capacity to represent the brain spatially but with about a six-second latency. EEG measures temporally down to the millisecond but can only assess brain patterns for the brain as a whole.

One observer said trying to figure out what is going on inside the brain is a little like trying to understand what is happening in a football game by standing outside the stadium. You can derive a few clues by listening to the dynamics of the cheering, but it is primarily guesswork.

Measuring your brain waves with an inexpensive EEG device became a bit of a fad in 2009 when Mattel released Mindflex, a game that used an EEG sensor to steer a ball through an obstacle course. It remains the best-selling consumer-based EEG device to date. Uncle Milton Industries that same year came up with Star Wars Force Trainer, a game designed to create the illusion of possessing the Force. Emotiv and similar companies continue to sell snazzy-looking EEG headwear to help with concentration for work and study and to support meditation. Users report that with some practice they can produce brain wave patterns that trigger the monitoring electronics, but it is difficult and error-prone—perhaps a little like playing golf with a toy tennis racket: you can get the golf ball to move, but it's awkward.

Among the most enthusiastic and successful of BCI researchers are Marcel Just and Tom Mitchell at Carnegie Mellon. They asked subjects to think about ten objects—five of them tools like a screwdriver or hammer and five of them dwellings, such as an igloo or castle. They then recorded and analyzed the activity in the subjects' brains as they concentrated on the assigned object. The research used the latest fMRI technology and assessed activity in about 3,000 little squares called voxels (about 5 millimeters square and containing perhaps a half-million neurons) and focused on a key subset of 120 voxels, which proved useful in distinguishing the assigned thought-objects. This is a technique sometimes called barefoot empiricism or black-box theorizing. Researchers simply correlate brain activity patterns with thinking about screwdrivers but without much understanding about how the correlation works. Professor Just speculates that when "you think of a screwdriver, you think about how you hold it, how you twist it, what it looks like, what you use it for," each of which may correspond to a set of locations in the brain. He did a dramatic demonstration for the television newsmagazine *60 Minutes* in which his system got ten out of ten guesses correct when the program's producer thought about various objects while inside an fMRI scanner as it "read her mind." Of course, that is ten out of ten prearranged and very specialized objects with the subject thinking intensely about a single object for an extended period of time. But it's a start. What is particularly important about Just and Mitchell's work, however, is that they find the same areas light up in response to the same object for every subject. So when the television producer volunteered, they used data from other subjects to base their guesses about which object was on her mind. Other researchers have found much more variability in matching neuronal activity and resultant behaviors across individual organisms. So, again, the research has a long way to go.

Some adventurous entrepreneurs like Elon Musk are not content to stand outside the stadium—they have more obtrusive plans. His start-up Neuralink based in San Francisco is planning to drill four 8-millimeter holes in subjects' skulls (pending FDA approval) and then insert threads that will pass neuronal data to an implant behind the ear, "as many as 3,072 electrodes per array distributed across 96 threads," according to a white paper. Musk's group, like others using surgical techniques, focus on patients with various serious impairments that justify the invasive technology. Musk characteristically promises great progress soon, "within a year," but the work remains proprietary and secretive, so it is difficult to evaluate.

The bottom line is that direct neurological communication, by which EI systems and portals could communicate directly with the brain, is a ways off—probably a matter of several decades. We await two fundamental breakthroughs: (1) finding a practical way to communicate with individual neurons or neuron groups without invasive surgery, and (2) having a better understanding of how the brain actually works. Because currently available BCI devices are awkward, intrusive, error-prone, and slow, we might be dismissive of the whole idea. That would be a big mistake. Direct neurological communication is likely, ultimately, to be an important part of evolutionary intelligence.

One way to think about it is as an inevitable step as computational intelligence has moved from a remote room-sized computer to the laptop and smartphone to the smart earbud. The smart earbud uses the traditional auditory channel. What's next? Currently cochlear implants for the hearing impaired translate aural information from the environment and electrically stimulate the auditory nerve through the cochlea in the inner ear. The amazing plasticity of the brain gradually learns to interpret these signals as sound as if from the vibration of the eardrum. Attempts to directly connect to the optic nerve, to date, have been less successful but are actively being pursued. These examples explore the brain's capacity to interpret neural electrical signals as sound and sight. They are likely to be the stepping stones to even more direct and holistic neural connection.

Imagine the following scenario. Today, your stomach growls with hunger and you look up from your work, utter the wake-up word "Alexa," then ask "what's for dinner?" And she kindly suggests some recipes. Tomorrow, your stomach growls with hunger and you look up from your work, utter the wake-up word "Alexa" followed by "what's for dinner?" And she senses the urgency in your voice and kindly suggests some quick-to-prepare recipes, thoughtfully based on what she knows is in your fridge. Ultimately, your stomach growls with hunger and you look up from your work and *think* the wake-up word "Alexa" followed by the thought "what's for dinner?" A wordless conversation ensues as you and your EI interface exchange views on the pluses and minuses of various options. It is the natural progression. There is nothing in the fundamentals of physics or biology to preclude direct wireless communication. Human ingenuity will figure it out. Dinner is decided. You're going out for Chinese. Nothing particularly superintelligent about that. Actually, you weren't really paying attention to all the pluses and minuses. You were just in a mood for Chinese. Some things won't change.

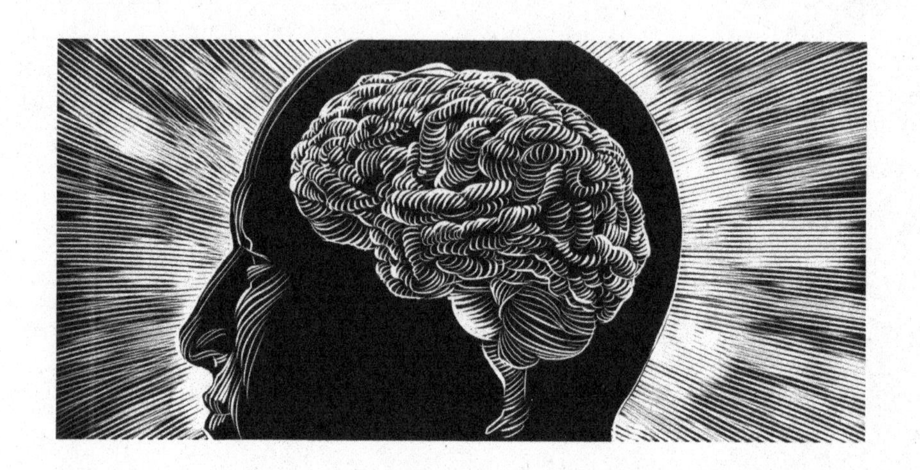

3 The Hard Part—Compensating for Human Cognitive Bias

We have asked: Evolutionary intelligence for what purpose?

We have answered: Counterpoise. Mediation. Moderation. Intercession. Compensation. Most importantly, systematic compensation for the increasingly well-understood biases and gaps in human cognition. We can use accumulated, evolved, time-tested artificial intelligence (AI) to advise us when we are systematically misinterpreting the cues in our environment. To inform us of otherwise unanticipated consequences. To correct predictable errors in our psychological estimates of probabilities. This is the hard part. This is what makes evolutionary intelligence (EI) potentially more than just another technical stepping stone.

In the abstract, compensatory EI sounds pretty straightforward. It is captured in the expression, "If I only knew then what I know now." When individuals or even institutions misunderstand their environment, misread social or market signals, underestimate probabilities, fail to consider alternatives, and ignore feedback (especially negative feedback), it would seem that some timely advice from a trusted source would be a big help. That's the basic idea.

Evolutionary Intelligence

Let's take a step back and work through what is implied by the concept of intelligence. Human intelligence is a complex and contested concept but can generally be taken to be the ability to accurately perceive information from the environment and to permit adaptive behavior, given an individual's goals. Culturally, we tend to associate intelligence with the accumulation of information, facility with language, and quick wittedness, but at its core intelligence is facility at matching means and ends. Thus, it is not surprising that various elaborations of "multiple" intelligences include social, spatial, and physical skills that are also useful coordinating means and ends. One group of psychologists, for example, emphasized that intelligence "is not merely book learning, a narrow academic skill, or test-taking smarts. Rather, it reflects a broader and deeper capability for comprehending our surroundings—'catching on,' 'making sense' of things, or 'figuring out' what to do."

Evolutionary intelligence, accordingly, is the technical facilitation of accurately perceiving our environment. So the timely provision of facts and figures is important, but that is less of a challenge. Think of the hubbub over IBM's Watson software outdoing two Jeopardy trivia champions in 2011. Looking up the fact that Helena is the capital city of Montana, looking up the weather forecast for tomorrow, finding a fact in a list of facts—the sorts of things Siri and other AI tools do well—is useful but not revolutionary in its impact.

Demonstrating the agility to acknowledge misperceptions when countervailing evidence begins to mount—that's more difficult. It is said that doing the same thing over and over again but expecting different results is akin to stupidity or even insanity. The axiom is attributed to Einstein. He may never have actually said that, but it is understandable that such an insight would be associated with his reputation. The popularity of the axiom reveals how clearly we see ourselves in it.

Why would an individual engage in the extra effort of evolutionary intelligence when primordial, self-interested gut instinct provides us behavioral guidance immediately, forcefully, and unbidden? The answer comes over time because we are rewarded as our accuracy in evaluating the environment improves. Think of the transition from the quill pen to the fountain pen to the ballpoint pen to the typewriter, electric typewriter, and then the word processor. Some of an older generation, perhaps set in their ways, couldn't be convinced that the awkward and expensive (potentially leaky) fountain pen was much of an improvement. Others stuck with their typewriters. But each generation of text-creating technology proved its worth over time and came to dominate how people wrote.

Early editions of the next generation of technology are likely to be expensive, awkwardly designed, and error-prone. They are more likely to be adopted by the young. But as the evidence of superiority accumulates, the inertia of habitual use slowly shifts.

Set aside for the moment the actual technical implementation of EI (visual, aural, tactile, etc.) and simply imagine the role of a calm, well-informed, discreet human-like adviser at your side.

Think of an interplay like the following:

Adviser: This looks like an important decision.

Self: Yes, very much.

Adviser: Do you have a strong gut instinct to pick choice A?

Self: Yeah, A.

Adviser: Why is that?

Self: I dunno, always picked A in the past.

Adviser: Let's run through the costs and benefits of A versus B.

Self: OK, good idea. This decision is important.

Adviser: Also, have you considered cost X and benefit Y?

Adviser: Have you considered options D and E?

The rhythm of such an exchange might remind you of the role of a therapist, financial adviser, primary care physician, a teacher, a coach, or a concerned friend or parent. In each case the advisory dynamic takes a similar form: "Wait a minute, let's clarify our goal here and our plan of how to achieve it." It turns out that such an advisory dynamic corresponds to the dual process model of human cognition that has its roots in the work of Harvard psychologist and philosopher William James in the late nineteenth century. The basic idea is that we tend to respond to our environment with (1) habitual, unconscious, instinctive, emotional reactions or (2) more deliberative and logical responses. More recently, Daniel Kahneman popularized this distinction as thinking fast or slow. So in this model of human behavior, all of us have an internalized advisory system perhaps too often ignored. The critically important signal may simply be "Wait a minute, let's think this through." One defining characteristic of EI, however, is real help in thinking it through.

Let's explore why help with thinking things through turns out to be the hard part.

The first mechanism is based on the premise that most human interactions are routinized and predictable in following familiar social scripts. As a result, identifying the dynamics of most "situations" is not necessarily an impossible task. Typical examples include the question-and-response process when using most tax-filing software programs, interviewing for a job, coordinating a social event, buying a car, and asking for a date.

The second mechanism is based on our characterization of EI as an advisory process, not a determining directive of human behavior. So, the human actor recognizes that a decision needs to be made. The advice, such as it is, may be forthcoming or explicitly asked for. It is part of a repeated interaction. The actor and adviser develop a rhythm and mutual understanding, and over time, the actor comes to acknowledge that the advisory resources offered are often useful and productive under certain circumstances. Take, for example, the driver and the advisory route suggested by a Waze-type GPS app. Waze suggests a route that gets you there three minutes sooner, but it involves numerous turns and a few edgy back roads. So you decline and opt for the simpler but slightly longer route. Waze learns from your preferences and adjusts its recommendations in later encounters. Humans evolve behavioral routines with each other. One comes to recognize and take advantage of another's expertise in Italian restaurants but not in sports betting and so on. The same will be true of our evolved relationship with the phenomenon of evolutionary intelligence.

As we address the hard part, it becomes increasingly evident that as EI remains at the very earliest stages of development, it may often be in error, irrelevant, or inappropriate. It will undoubtedly lead to unintended consequences. Better to think through these intended and potentially unintended consequences before they happen rather than afterward.

Human Cognitive Biases

Let's take a look at the diverse variety of demonstrable systematic distortions in human cognition. As we noted previously, in its characteristic impulse for comprehensiveness, Wikipedia has put together a list of over 200 specific and largely empirically demonstrable "cognitive biases." Here are a few highlights.

Selective sample of common cognitive distortions	
Confirmation bias	The tendency to search for, interpret, focus on, and remember information in a way that confirms one's preconceptions.
Commitment bias	The tendency of people to support their past ideas and decisions even when presented with evidence that they are wrong.
Mood-congruent memory bias	The improved recall of information congruent with one's current mood.

Selective sample of common cognitive distortions	
Hot-hand fallacy	The belief that a person who has experienced success with a random event has a greater chance of further success in additional attempts.
Availability heuristic	The tendency to overestimate the likelihood of events with greater availability in memory, which can be influenced by how recent the memories are or how unusual or emotionally charged they may be.
Zeigarnik effect	Uncompleted or interrupted tasks are remembered better than completed ones.
Dunning–Kruger effect	People who are least competent at a task often incorrectly rate themselves as high performers.
Conjunction fallacy	Errors in categorization of human attributes.

Predictable distortions of perception in the evolved human cognitive system represent a serious challenge and, I suggest, perhaps even an existential challenge. Confirmation and commitment biases, for example, are so obviously important and counterproductive that professional training in medical diagnosis, law enforcement, and business administration include exercises to try to overcome these constraints on our capacity to fairly evaluate evidence. Our capacity to "explain away" contrary evidence by means of ad hoc hypotheses is well established. Systemic cognitive biases, unfortunately, are not always corrected by group process or collective wisdom. Sometimes it is quite the reverse. They may be exaggerated by collective practices; notably, crowd or mob-like behaviors as misperceptions may be reinforced and energized by crowd dynamics.

The mood-congruence bias is evident when you are more likely to remember something that's happened to you if the memory matches your current emotional state. Sad memories come more quickly to mind when we are feeling sad. These phenomena emphasize how fragile and unreliable our memories are. Our brain does not store memories like files in a filing cabinet. It is a fraught dynamic process as we remember remembering past events, since the two mix together often with creative embellishments and deletions.

There are numerous well-known cognitive distortions associated with gambling. We pay close attention to winning (and losing) streaks, thus the hot-hand phenomenon. Perhaps you knew that Joe DiMaggio has held the Major League Baseball record for consecutive games with at least one base hit. Did you know his streak of 56 games in 1941 shattered the previous record of 45 games set by Willie Keeler in 1897? Many baseball fans know this well and will be happy to explain the details to you. Behind this particular obsession with sports statistics is the obviously erroneous belief that a previous random event will affect subsequent ones. If you flip six heads in a row, wouldn't you expect the odds of tails on the next flip would increase? You are not alone. And, of course, you're wrong. Celebrating notable historical winning streaks is harmless. Betting on future ones is not. Six heads in a row does not make a heads in the next flip any less likely.

The availability heuristic is parent to perhaps dozens of predictable distortions in human judgment. We make decisions drawing on the first concrete example that comes to mind. Thus, there are many illustrations of recency effects, whereby most recently received facts or figures are most influential on our decision making. What would you say? Are English words starting with the letter "k" more common than words in which "k" is the third letter? (They aren't.) But it seems to make sense because we are better at recalling words that start with k. Word frequencies illustrate the phenomenon. This is, however, a serious enough distortion to generate demonstrable biases in financial investments. Two Israeli analysts found that investors were more likely to accept upgrade recommendations for specific stocks when the general stock market index was positive. The reverse was true as well for downgrade recommendations. They identified this investment bias as clear evidence of the availability heuristic.

The Zeigarnik effect sort of makes evolutionary sense. When a task is completed we want to set aside the details and move on to the next task at hand. Given the limitations of human memory, it would seem to be the cranial equivalent of keeping a tidy work desk.

It is said that "ignorance more frequently begets confidence than does knowledge." The axiom is attributed to Charles Darwin. It turns out that not only was he right, but he identified a demonstrable bias in our inability to evaluate our own skills and shortcomings. The less we know about a field of endeavor, the better our estimates of our mastery of the field. A particularly ironic pattern of misperception is known as the Dunning–Kruger effect. These social psychologists studied students' self-perceptions in the areas of humor, grammar, and logic and found consistent evidence that ignorance breeds confidence in self-appraisal. The relevance of this problematic dynamic of distortion for the workplace is obvious. The least informed may have the loudest voices in collective decisions.

Then there is the conjunction fallacy. Linda is 31 years old, single, outspoken, and very bright. She majored in philosophy. As a student, she was deeply concerned with issues of discrimination and social justice and also participated in antinuclear demonstrations.

Which is more probable?

A. Linda is a bank teller.

B. Linda is a bank teller and is active in the feminist movement.

Most of us pick B. It makes intuitive sense. But by definition, unless all bank tellers are feminists, the number of bank tellers who are feminists is going to be a smaller number than the total number of bank tellers, so in terms of strict probability, A is the better answer. The common error of association of human attributes is known as the conjunction fallacy. The probability of two events occurring in conjunction is always less than or equal to the probability of either one occurring alone. I know, I know. You still think Linda is likely to be a feminist. But that wasn't the question.

Our diverse collection of demonstrable and systematic errors in human judgment does make a case about the prevalence of cognitive bias. The listed patterns of distorted thinking is amusing, perhaps distractingly so. Such errors would appear to be the stuff of parlor games and innocent mistakes. But cognitive bias is serious business.

Is there a common element in these hardwired propensities for us to misperceive our environment? I believe there is. It derives from the fact that our judgments about our own behaviors are systematically different from our judgments of others. Although such a mindset may have served us well in the evolution of familial and tribal existence as we hunted and gathered and survived in existential competition with rival tribes nearby, it does not serve us well in our modern intertwined global existence. Psychologists call it the fundamental attribution bias. We will develop this theme in more detail ahead. Dealing with this deeply engrained pattern of perception may turn out to be the hardest part of the hard part.

I have been characterizing EI in rather concrete terms as a sort of human adviser. It is a useful way to communicate the essence of the concept. But hopefully the concreteness of the adviser metaphor does not become a distraction. I'm not proposing silicon shoulder angel (or an angel on one shoulder and a devil on the other.) I'm not proposing a Wi-Fi Jiminy Cricket. We have speculated about how helpful information could be technically manifested in visual, aural, tactile, and neuro-direct modalities. In some cases helpful advisories will take the form they currently do in the electronic environment as text-based suggestions on a screen as one fills out a form or perhaps as a well-understood warning symbol or sound from a computer monitor. The modality is not critical. What is critical is the mechanism of how EI might actually serve to improve human decision making.

How We Think

Imagine that you are a newly admitted first-year student at a prestigious college and have decided to major in psychology. You think you are pretty smart and quite proud about getting into a good school. Pretty smart? The Introduction to Social Psychology course conveys a very different message—humans routinely misconstrue signals from their environment, are easily fooled and distracted, forget quickly, and are overly responsive to perceived social pressures. Maybe not so smart. There may be some room for EI to help. One could review the rather disheartening list of classical experiments routinely featured in most introductory social psychology curricula.

Solomon Asch's Conformity Experiments

The studies took place at Swarthmore in 1951. A naive research participant in a purported study of visual perception is asked to judge whether lines on a display card are longer or shorter than a reference display card. There are, however, seven additional participants in the room who are experimental confederates making the same judgments publicly and who have been coached to give incorrect answers together at certain intervals. The actual answers about which lines are longer or shorter are clear. Independent tests without confederates have results that are 99 percent accurate. With seven "votes" at odds with your own judgment, you might well bend to the perceived social pressure. Only 25 percent of the naive participants consistently resisted the social pressure to misjudge the stimuli. We are easily influenced.

The Stanford Bias Experiments

Lee Ross and his associates at Stanford conducted numerous experiments that demonstrate the difficulty people have in responding to evidence that conflicts with their strongly felt opinions. In one study of opinions on capital punishment conducted in 1979, students who favored capital punishment were shown detailed research that compared otherwise comparable states with and without capital punishment. The results of the studies challenged the view that capital punishment had a deterrent effect on serious crime. One would expect in the face of such conflicting evidence that supporters of capital punishment might moderate their views, but the result was just the opposite: Their support of capital punishment increased, dramatically. They discounted and dismissed the contrary findings. And, as one might expect, the results were the same but in the opposite direction for students who opposed capital punishment.

Miller's Magic Number Seven, Plus or Minus Two

Harvard's George Miller published what would become a classic article on human cognitive capacity that focused on our limited capacity to discriminate among stimuli. Working with a diverse set of experimental situations, Miller demonstrated that when the number of objects exceed about seven, we get confused and disoriented. He described the various "chunking" heuristics we humans have developed (grouping similar objects) to keep us from overloading and distorting or cognitive capacities.

Kahneman and Tversky's Demonstration of Systematic Human Cognitive Errors

As noted previously in chapter 1, this work is perhaps most famous, meriting a Nobel Prize and stimulating work in multiple academic fields, including behavioral economics, game theory, and decision theory. The list of demonstrable and presumably hardwired errors of human judgment they demonstrated is long. Let's consider a few.

The framing effect refers to people's tendency to be influenced by the way a problem is posed or a question is asked. Subjects were asked to imagine that the United States was preparing for an outbreak of an unusual disease expected to kill 600 people and that two alternative programs for combating the disease had been proposed.

For one group, the choices were framed in terms of gains: How many people would be saved?

If program A is adopted, 200 people will be saved.

If program B is adopted, there is a one-third probability that 600 people will be saved and a two-thirds probability that no people will be saved. In this case, the majority of subjects pick the first option, program A.

For another group, the choices were framed in terms of losses: How many people would die?

If program C is adopted, 400 people will die.

If program D is adopted, there is one-third probability that nobody will die and two-thirds probability that 600 people will die.

In this case, the vast majority of volunteers were willing to take a risk and selected program D.

The options in both cases are, of course, the same, just framed differently in terms of gains or losses. The first program would save 200 people and lose 400. The second program offered a one-in-three chance that everyone would live and a two-in-three chance that everyone would die. Framing the alternatives in terms of lives saved or lives lost is what made the difference. When choices are framed in terms of gains, people often become risk-averse, whereas when choices are framed in terms of losses, people often became more willing to take risks.

WYSIATI—what you see is all there is. Since humans are inveterate story-makers, they are inclined to build a theory for what causes what, using whatever data is available. Kahneman characterizes the human brain as a "machine for jumping to conclusions." One example of this phenomenon is our reaction when meeting a stranger. We take only a few seconds to build up an impression, perhaps deciding whether the individual is friendly or hostile or dominant. Of course, we have only fragmentary information obtained from their posture, demeanor, and facial expression. We develop a coherent storyline on fragmentary impressions. We feel the need for a story, so we invent one. It is not a recipe for successful human interaction.

The impression our potentially overwhelmed first-year college student may reach in confronting all these provocative studies may not be that we are irretrievably foolish but rather that we rely on emotions too often instead of putting in the effort to think things through. And that's a good point.

Princeton's Susan Fiske characterizes our dual modes of automatic instinctual and more effortful controlled and deliberative thinking (corresponding to Kahneman's fast and slow) this way:

- Automatic processes make social thinking efficient.
- Controlled processes make social thinking flexible.
- Motivations and situations influence which of these modes operates.

The term of art in psychology for the human trait of efficiency is to characterize us as "cognitive misers." That makes it almost sound reasonable. We are perfectly capable of thinking things through and weighing alternatives but need extra motivation to move above our stereotypes and impulses. Bottom line, we are lazy thinkers.

Although it is tempting to continue building an expansive list of hundreds of human cognitive shortcomings, it may be most useful to organize them—that is, chunk them together in coherent clusters and in such a way that we can review how evolutionary intelligence has a prospect of making a compensatory difference. Given George Miller's sound advice, perhaps seven would be a good working number of collective categories. When you put it together in a single sentence, it is quite an indictment of the human condition.

Humans are lazy, uninformed, biased, inattentive to feedback, isolated, and emotional. But no problem. As they say in Silicon Valley, there will be an app for that.

Here's the cognitive indictment in tabular form. We'll walk through each of the seven categories. As we do, you will no doubt make note that what I am describing as compensatory mechanisms are clearly beyond the current operational capacities of artificial intelligence systems. Duly noted. But that is the largely the point—to explore how the next generation of artificial intelligence will not just do what it does now faster, but that it will be able to undertake entirely new functions in the service of its human creators. Please note also, as I proffer short conversational scenarios as a very smart and all-knowing future Siri signals to you, that this invented conversation is just a primitive representation of what in time will be a more abbreviated exchange of signals and abbreviations as you and your digital resource get to know and grow comfortable with each other.

Seven mechanisms of compensatory evolutionary intelligence

Category	Cognitive pattern	Evolutionary intelligence
Lazy thinkers	Rely on unthinking habitual responses	Motivate deliberative response
Uniformed thinkers	Have limited information in available memory	Provide timely information
Biased observers	Selectively attentive to environment	Can provide alert function
Biased evaluators	Misjudge cues in environment	Provide potential perspective of the "other"
Inattentive to feedback	Misjudge responses to our behavior	Provide behavioral scenarios
Isolated thinkers	Limited ability to access collective resources	Engage collective evaluations
Emotional thinkers	Influenced by transitional moods	Monitor corporeal responses

Lazy Thinkers—Connecting Systems 1 and 2

You may be familiar with this classic little puzzle. A bat and ball together cost $1.10. The bat costs $1 more than the ball. How much does the ball cost?

You really want to answer 10 cents. It comes to mind quickly. It sort of seems about right. You don't actually want to run through any arithmetic to see if that really adds up. Unfortunately, it doesn't add up. If the ball costs 10 cents and the bat costs one dollar more, the bat would cost $1.10 and the two together would cost $1.20. The correct response is 5 cents for the ball and $1.05 for the bat.

So are you a particularly lazy thinker? Not really. The puzzle is designed to catch you off guard. What you are is a typical human thinker who thinks with reasonable efficiency. Social psychologists refer to this a dual process thinking—the dynamic interaction between two physiologically independent systems in the brain.

As Fiske and Taylor explain in their classic text on social cognition:

> System 1 versus System 2 contrasts intuition versus reasoning. The intuitive side is holistic, rapid, effortless, parallel, affective, associative, crude, and slow-learning. The rational reasoning side is analytic, slow, effortful, serial, neutral, logical, differentiated, flexible, and fast-learning. The intuitive, associative mode relies on the slow-learning but then rapidly responding form of memory, which concentrates on consistencies. In contrast, the rational, fast-binding, rule-based system acquires detailed new memories quickly, focusing on novel and inconsistent cues for subsequent consideration in deliberate processes. The two-system contrast suggests that each is suited to distinct forms of learning and reacting. This powerful theme drives much social cognition work as psychological scientists of every stripe document distinct ways people make sense of their social worlds.

They summarize the distinction in tabular form.

Automatic System 1	Controlled System 2
Rapid	Slow
Intuitive	Rational
Categorical	Analytic
Effortless	Effortful
Slow-learning	Fast-learning
Affective	Flexible
Crude	Novel
Reflexive	Reflective
Rigid	Logical

The multiple locations of these interacting systems in the brain reflect a complex pattern, still contested among neurophysiologists, but that might be roughly summarized as shown in the following figure.

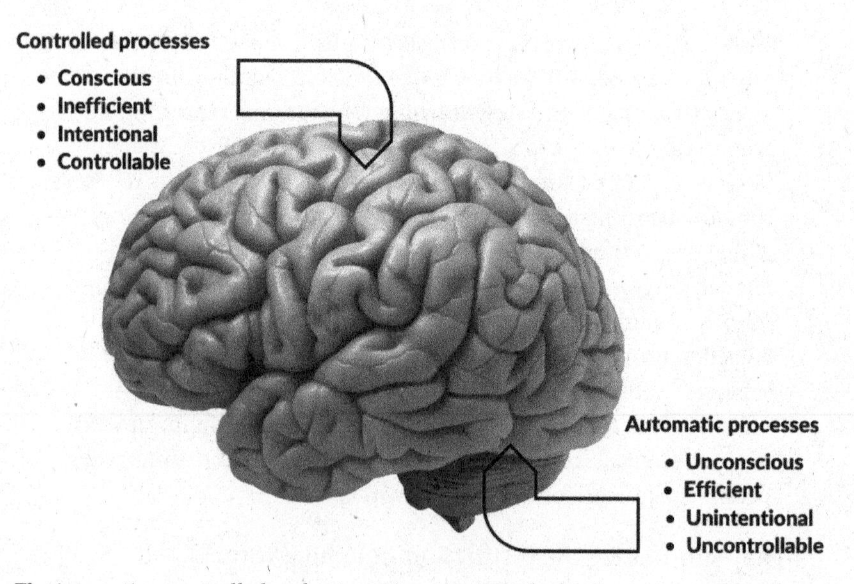

The interacting controlled and automatic systems in the brain

The neural correlates of consciousness and control tend to be found in the outer, upper (more recently evolved) shells of the cerebral cortex while automatic "unthinking" processes are lower down around the amygdala and brain stem, referred to by us amateur observers as the original lizard brain of our animal ancestors. Accordingly, we tend to characterize the automatic, instinctual processes as more primitive and to be avoided. But psychologists remind us that rather than lazy thinkers, we should be characterized as appropriately efficient thinkers not wasting time on what should be routine and automated behavioral responses. If you had to think about and reevaluate each individual movement as you drive a car it would be paralyzing (and probably cause an accident.)

Kahneman has popularized this element of social psychological theory in his influential book *Thinking Fast and Slow*, published in 2011. He warns against minimizing the importance and practical significance of System 1 thinking:

> System 1 is indeed the origin of much that we do wrong, but it is also the origin of most of what we do right—which is most of what we do. Our thoughts and actions are routinely guided by System 1 and generally are on the mark. One of the marvels is the rich and detailed model of our world that is maintained in associative memory: it distinguishes surprising from normal events in a fraction of a second, immediately generates an idea of what was expected instead of a surprise, and automatically searches for some causal interpretation of surprises and of events as they take place.

As he goes on in his explanation, here's where evolutionary intelligence comes in:

> The way to block errors that originate in System 1 is simple in principle: recognize the signs that you are in a cognitive minefield, slow down, and ask for reinforcement from System 2. . . . Unfortunately, this sensible procedure is least likely to be applied when it is needed most. We would all like to have a warning bell that rings loudly whenever we are about to make a serious error, but no such bell is available, and cognitive illusions are generally more

> difficult to recognize than perceptual illusions. The voice of reason may be much fainter than the loud and clear voice of an erroneous intuition, and questioning your intuitions is unpleasant when you face the stress of a big decision. More doubt is the last thing you want when you are in trouble. The upshot is that it is much easier to identify a minefield when you observe others wandering into it than when you are about to do so.

A warning bell that rings loudly whenever we are about to make a serious error. Sounds like a great idea. An even better idea would be reminders of the last time a similar mistake was made and perhaps a few scenarios of alternative approaches to the issue at hand.

Uninformed Thinkers—Timely Information Retrieval

"Jaywalking" was a prerecorded segment on *The Tonight Show with Jay Leno* in the early 1990s through 2014. On walks down Hollywood Boulevard and Melrose Avenue, Leno asks bystanders, many of them tourists, various questions about history and public affairs. It was a celebrated demonstration of how little most Americans know. To make it obvious that these random pedestrians should know the answers, he would often ask these self-reported American citizens questions from the actual U.S. Citizenship Test. The results are often (unfortunately) hilarious as, for example, one woman said that the American Revolution was fought against France and that Abraham Lincoln was the first president. Others struggled with questions such as "What color is the White House?" and "In what country is the Panama Canal located?" The point in this case was humor and the capacity of late-night viewers to go to bed feeling smugly superior. But uninformed thinkers are problematic.

Political scientists point out that democracy based on misinformation and a lack of information is a recipe for calamity. Ilya Somin, for example, points out that during the 2014 election "only 38 percent of Americans knew that the Republicans controlled the

House of Representatives at the time, and the same number knew that the Democrats had a majority in the Senate. Not knowing which party controls these institutions makes it difficult for voters to assign credit or blame for their performance." He notes that only about a third of respondents could correctly identify the unemployment rate; only one in five knew the federal government spends more on Social Security than interest on debt and foreign aid; and (my favorite), 35 percent believed that Karl Marx's dictum "From each according to his ability, to each according to his need" is in the Constitution.

Perhaps one could politely excuse a disinterest in the textbookish facts and figures of public affairs as an overly academic domain. One could argue the fields of personal health, finance, and consumer goods should fare better. They don't.

- An analysis of survey findings on Americans' understanding of basic information about biology and health revealed that just over half knew that antibiotics kill bacteria but not viruses or knew whether the father's or mother's genetic contribution determines the child's gender.
- American researchers found that public knowledge about dietary fats and cholesterol was quite poor: only three of the 11 questions asked in 1988 were answered correctly by a majority of consumers.
- In the United Kingdom, the Health Education Monitoring Survey in the 1990s found that only 16 percent of respondents could come up with as many as three out of four of the core widely publicized recommendations for a healthy diet.
- Only 12 percent of adults in the United States have a high level of health literacy, according to the National Assessment of Adult Literacy. In other words, nearly nine out of ten adults lack the skills needed to fully manage their health care and prevent disease. Low literacy has been linked to poor health outcomes,

higher rates of hospitalization, less use of preventive services, minimal prescription and care plan adherence, and death.

- A Standard & Poor's survey found just 57 percent of U.S. adults to be financially literate with even a basic understanding of financial concepts such as inflation and risk diversification in investment (worldwide among adults, the survey found 33 percent financial literacy).
- A compilation of 1,200 online surveys of U.S. adult citizens found that nearly half reported that they are extremely or very financially literate. However, when asked to take a financial quiz, fewer than half passed, and only 6 percent scored an "A" grade of 90 percent or better.

George Miller's magic number seven reminds us that it is just physiologically reasonable to assume that we can only manage a certain number of facts at a time. Forgetting details keeps us from being simply overwhelmed by the complexity of our modern lives. Providing timely facts and figures is perhaps the easiest and most straightforward exemplar of technologically enhanced EI. It is what Siri and her compatriots are already pretty good at—the weather, the traffic, appointment reminders, shopping lists, calories per serving, and the actual cost of an item given accumulated interest.

A few timely facts available to a consumer could indeed come in handy in certain marketplace scenarios. The quintessential case, of course, is comparative prices. With current technology, one can whip out a smartphone and scan a barcode of an item for a possible price comparison. That's a start. In the future, the process will be more automatic. Consumers will be in a position to set price difference intervals and ease-of-alternative-purchase thresholds or minimum items costs. The broad use of these technologies will have (from the consumer's point of view) a major impact on retail dynamics.

With current technologies, if you want a weather forecast you ask for it. In the future, one could imagine it will be provided as you reach for your jacket in the closet. But anticipating your interest is perhaps less impactful than providing information you didn't realize you might need. In the early stages of these ongoing developments, it will be the case that the system provides information that you didn't think you needed—and, in fact, you didn't. But over time and with sophisticated feedback loops in the process, the categories of information that make a demonstrable positive difference, personalized to your circumstances, will become clearer.

- It's dusk. Be on the lookout for jaywalking pedestrians as you drive this stretch of road.
- This item is actually tax-deductible.
- That burger is actually about 1,550 calories. Try the veggie burger without cheese. It's highly rated as very tasty, and you reported you liked it last time.
- The last time you saw Aunt Martha was Thanksgiving three years ago. Remember, she is hard of hearing; talk slowly but don't yell.
- Based on recent history, the chances you will pay off that payday loan on time are in the range of 12–15 percent.
- Your pin for this account is 7025.
- The odds of winning the $350 million Powerball drawing are roughly one in 292 million. You would have a reasonable chance of winning if you bet every day for the next 800,000 years. The odds you will be hit by lightning this year are over 400 times greater than winning this Powerball drawing.
- That broker has been recommending hot stock tips to you for years; he works on commission. If you invested in the last four recommendations he made, you'd be down $352,000.
- You are eligible for a new college-loan forgiveness program.
- It's time for your annual physical and teeth cleaning.
- You can get there faster and cheaper by renting a car and driving rather than flying.

Why is it when I look over a list like this I think of Mr. Spock from *Star Trek*? Spock, the iconically well-informed and seemingly emotionless observer, usually had an extremely reasonable suggestion or two. Maybe it is not such a bad idea to have him around making suggestions from time to time.

Biased observers and biased evaluators are centrally critical dynamics in the EI story. But they are best addressed separately. The dynamic of the biased observer refers to selective attention leading to missing information; the biased evaluator refers to the potential misinterpretation of a signal that has reached the threshold of our attention.

Biased Observers—Alert Cues

The class instructor is going to show a brief video as an attention-concentration exercise. The video depicts six students bouncing basketballs back and forth; three students are in white shirts and three in black. The class's attention assignment is to count the number of the tossed balls but only those between the students dressed in white. About 30 seconds into the video recording, a confederate in a gorilla suit casually walks in among the players. The gorilla looks at the camera, does some chest thumping, and then saunters away while the students continued passing the basketballs.

The instructor asks the class for their count of basketball passes. Most observers come pretty close to the actual number of 15. Then the instructor asks how many saw the gorilla. Usually about half miss the gorilla entirely because they were obediently following directions and, in doing so, demonstrating the impressive focusing power of human selective attention. In a demonstration to 400 students in an auditorium in London, only 10 percent noticed the gorilla.

Selectivity is essential because the brain couldn't possible process all the many thousands of sensory signals available at any moment.

In one study of human vision, the researchers concluded that less than one percent of the visual input data (at around one megabyte per second) enters the processing bottleneck of the brain. We see what we want to see and what we expect to see. This lies at the core of all the confirmation-bias-type errors in human judgment that Kahneman and Tversky famously demonstrated. It is not a new observation.

Back in 1922, Walter Lippmann provided a telling case study in his influential book *Public Opinion*. Lippmann takes some care to describe a study conducted a few years before in Germany in which a group of psychologists, unbeknownst to them, witnessed a brief and carefully choreographed scuffle among a group of actors who rushed in and then out of the room. This group of 40 presumably well-trained observers were immediately asked to write down in detail what they had just witnessed. Lippmann proceeds to describe the analysis of their "eyewitness" reports:

> Only one had less than 20% of mistakes in regard to the principal facts; fourteen had 20% to 40% of mistakes; twelve from 40% to 50%; thirteen more than 50%. Moreover in twenty-four accounts 10% of the details were pure inventions and this proportion was exceeded in ten accounts and diminished in six. Briefly a quarter of the accounts were false. . . . The ten false reports may then be relegated to the category of tales and legends; twenty-four accounts are half legendary, and six have a value approximating to exact evidence. Thus out of forty trained observers writing a responsible account of a scene that had just happened before more than a majority saw a scene that had not taken place.

Like missing information, the dynamic of drawing attention to potentially important signals in the environment would normally be welcomed by the individual.

- Beware the stop sign on the right partially hidden by that bush.
- Those are rain clouds; the chance of rain in the next ten minutes is 86 percent.

- The traffic light ahead is about to turn red.
- The caller ID on this incoming call is your accountant.
- The individual sending you this email has an extensive criminal record.
- Traffic to the airport is unusually heavy; you'll never make it. Better to rebook.
- This speech therapist is rated very highly by clients. The therapist's rates, however, are 15 percent higher than the average for this area.
- The current medical diagnostics from your smart watch monitor are very serious and strongly suggest you get to the ER immediately. Directions attached.

As biased observers, we miss many cues in our environment. As biased evaluators, we are looking directly at the evidence and misunderstanding it.

Biased Evaluators—The Hardest Part of the Hard Part

I'm not sure why, but we humans seem to derive pleasure from the fact that we get things wrong so often. Misunderstandings are central to the structure of humor. We laugh when someone misunderstands. Think of how often a misunderstanding is the engine of a stand-up routine or a situation comedy narrative—Abbott and Costello's "Who's on First?" routine, Lucy and Desi, *Seinfeld*. When a child misunderstands, it strikes us as cute.

Perhaps such an assertion needs to be qualified. We derive pleasure in observing someone else's misunderstanding, not our own. And that self–other asymmetry lies at the core of what we have come to understand about many of the hardwired biases in human cognition. We have a very different schema for evaluating our own behavior and that of others.

Experimental Demonstrations of the Fundamental Attribution Error

"Participants read a description of two students who were asked to participate in a quiz game. Two students were described who were randomly assigned the role of being either the quizmaster (who generated a number of questions based on his idiosyncratic knowledge to pose to the contestant) or the contestant. The story described the contestant as correctly answering only one of the five questions presented by the quizmaster. If perceivers take into account the situation—that is, that the quizmaster and contestant roles were randomly assigned—they should realize that the quizmaster would be likely to have fared just as poorly as the contestant if their roles had been reversed." But sure enough students tended to rate the contestant as less "smart" than the quizmaster. If the students had role-played being the contestant, no doubt they would have asked: "Wait a minute, the other guy got to think up the questions."

A good example of this asymmetry is what has come to be called the *fundamental attribution error*. Put briefly, we tend to blame external situational factors when we fail and credit our personal capacities when we succeed, and we do just the reverse for others. It wasn't my fault, but in your case, I blame your personality and disposition. There are numerous classic psychological experiments that illustrate this persistent cognitive distortion. It is part of a broad array of attributional distortions when we try to assess causes and effects. It is our nature to keep building theories of what is causing what. But we are not good at it. We have more empathy for ourselves and people like us and skepticism about others. Such cognitive predispositions, as we have noted, may have served us well through prehistoric tribal competition for scarce resources. But in the age of global trade, instantaneous worldwide communication, and nuclear warheads. . . . maybe not so much. Time to progress to the next stage of evolutionary intelligence.

The key notion here is a hardwired asymmetry in how each of us perceives our environment. By definition, it cannot always be the case that the self is situationally constrained and the observed other is not. That would quickly become evident if the self and the other reversed roles. This notion of taking the perspective of the other will play an important role in EI.

Drawing on the work of psychologist Ziva Kunda, researchers have labeled this domain of psychological dynamics "motivated reasoning." As Jost and Krochik summarize the field:

> The study of motivated reasoning—that is, the ways in which individuals, because of psychological needs, goals, and desires that shape information processing, reach conclusions that they (on some level) wish to reach rather than ones demanded by adherence to logic or evidence—may have originated with Sigmund Freud, but it has become something of a cottage industry in psychology. Several decades of research in experimental psychology and, more recently, political science demonstrate that people frequently process information selectively to reach desired conclusions while avoiding information that, objectively speaking, would lend credence to alternative points of view. . . . As Stephen Colbert put it, there may be important differences between "those who think with their head, and those who know with their heart."

Kunda emphasized that although motivated reasoning is ubiquitous, it is not inevitable. Experiments show that when individuals expect that they will have to explain their decisions or that they will be evaluated, the influence of motivated distortions in reasoning declines. She labeled this condition "accuracy motivation." We face many decisions routinely—many of them trifling and trivial, such as selecting a suit for the day, choosing lunch options, picking a movie. Since comparing our preconceptions against the available independent evidence involves energy and effort, it may be harmless to engage and simply go with our gut feelings and impulse for the moment. But for life's more important decisions such as financial, career, and health choices and personal or professional negotiations with others, we may want to put a bit more effort into accuracy motivation in our attributions.

We may be well aware of the fundamental attribution error and its many corollaries. We joke about it. For instance, I'm a Red Sox fan, so don't ask me to acknowledge that the Yankees ever did anything admirable—and so forth. We recognize the confirmation bias, although we may be dramatically more attuned to its decision-distorting power among others rather than ourselves. The key to this aspect of EI is putting that self-awareness to work.

The classic manifestation of mutual perception or perhaps mutual misperception is a two-person game, and the classic game of this tradition is the *prisoner's dilemma.* It turns out, after many thousands of trial runs of the classic dilemma, there actually is an optimal strategy for dealing with others when you are not so sure about whether trust is warranted.

> Two prisoners are in separate cells, unable to communicate, accused of a crime. The authorities offer the prisoners a deal. They can testify to the other's participation in the crime, with the following conditions:
>
> If both prisoners inform on the other, each will serve a three-year prison term.
>
> If one prisoner informs on the other and the other stays silent, the accused will serve five years and the accuser zero.
>
> If both prisoners stay silent, each will serve one year in prison.

In game theoretic terms, choosing to stay silent is "cooperating" (in this case, cooperating with the other prisoner) while accusing the other is "defecting."

What would you do? How do you evaluate the other's trustworthiness to stay silent? It would seem mutual cooperation is the best deal, but only if the other is trustworthy. Getting off scot-free by snitching on the other is an attractive option. But the cost of cooperating while the other defects is pretty high. A dilemma.

The payoff matrix: Classic prisoner's dilemma

		Player Two	
		Cooperate	Defect
Player One	Cooperate	1, 1	5, 0
	Defect	0, 5	3, 3

(Player One prison sentence in years, Player Two prison sentence in years)

The iterated prisoner's dilemma is where it gets interesting. The game is played multiple times with the same payoff matrix, and each player learns from observing the other player's behavior in the previous games. If both cooperate each time without exception, it appears to be optionally rational, if a bit boring. But what if Player One defects, every once in a while, randomly, just to push up their own score? What should Player Two do? See what I mean?

Robert Axelrod's *The Evolution of Cooperation* published in 1984 made a case study of the iterated game. Axelrod organized a public tournament in which participant pairs would test consistent strategies over many iterations. Some players tested largely cooperative and others largely defecting game plans, and still others developed complex algorithms mixing hostility and forgiveness.

The tournament revealed that greedy strategies tended to do very poorly in the long run while more altruistic strategies did better. Axelrod interpreted this result as a possible mechanism for the evolution, by natural selection, of altruistic behavior from mechanisms that are initially purely selfish. But, it turns out, there was a clear winning strategy for the long term.

The strategy was dubbed "tit for tat"—cooperate on the first iteration of the game; after that, simply follow what the opponent did on the previous move. Sort of do unto others what they just did to you. There are some slightly more complicated variations of this tit-for-tat strategy that add occasional extra forgiveness rounds of cooperation to prevent an endless cycle of defections. But the simplicity and elegance of the winning strategy represented a striking finding in Axelrod's research. It turns out this game dilemma is surprisingly similar to many real-world scenarios and thus highly applicable. Studies have found parallels in business competition, sales negotiations, party-based legislative competition, and even domestic negotiations.

I have dubbed this the hardest part of the hard part because it is challenging for humans to successfully take the perspective of the other. When circumstances appear to be zero-sum, in which some advantage to the other appears to be a loss for us, we snap into fundamental attribution error mode. The other player is nasty, unfair, selfish, and probably cheating. Me, I'm just trying to put food on the table to feed my hungry children. It would appear that the other player never has children or gets hungry.

When economic disputes get serious (and likewise for domestic disputes), our evolved cultures have designated independent authoritative tribal, judicial, or religious institutions to facilitate negotiation. The idea behind EI is that this becomes more generalized and routinized, and its obvious benefits become more accessible and informal.

- Yes, you are convinced your tennis serve was in bounds; unfortunately, the electronic record indicates unambiguously that it was out of bounds.
- Your potential opponent has a long history of litigiousness, much of it unsuccessful but nonetheless unrelenting; it might be prudent to take a small loss in this case.
- Although tipping is not common in this country, the tourist industry recognizes Americans and has come to expect them to tip generously.
- You won $300 at this casino on the last visit. You have positive feelings toward this place. That is a recency effect (i.e., a priming, top-of-mind effect). The long-term odds of winning at this casino are a negative $79 per hour (a negative $127 per hour if travel and lodging costs are included.) Don't forget the $680 you lost here last June.

Inattentive to Feedback—Scenario Modeling

Recall the Zeigarnik effect from our list of cognitive biases. People tend to remember unfinished or interrupted tasks better than completed ones. It turns out the brain keeps tabs on unfinished tasks and then sort of "cleans out" short-term memory when the task is done to make room for the next one. The phenomenon is named after Russian psychologist Bluma Zeigarnik. As the story is told, a colleague had made note of a waiter's amazing capacity to remember unpaid orders while having no recollection of them after the diners had left. She decided to test and demonstrate the generality and consistency of the phenomenon in a series of carefully crafted experiments in the 1920s. It would seem to be evolutionarily useful in keeping one's head clear of unnecessary details. (My customary way of thinking of this is to note how critically important it is to remember the number of your hotel room while you're there and how utterly irrelevant it is after you have moved on.) Where it may become evolutionarily counterproductive, however, is when important feedback in response to our behaviors is not immediate and we are no longer attentive to keeping our behavior informed by positive or, importantly, potentially negative feedback.

Part of inattentiveness to feedback can be understood as largely an extension of the lazy thinker. Thinking about our environment takes effort. It is easiest to coast unthinkingly, following well-established routines. Psychologists respond to this with a variety of exercises under the rubric of mindfulness. The literature following Canadian psychiatry professor Scott Bishop has converged on a two-component model of mindfulness. The first component involves the self-regulation of attention so that it is maintained on immediate experience, thereby allowing for increased recognition of mental events in the present moment. The second component involves adopting a particular orientation toward one's experiences in the present moment, an orientation that is characterized by curiosity, openness, and acceptance.

As stated in this mindfulness model:

> Much of cognition occurs in the service of goals. We are constantly engaged in a process of comparing what is with what is desired,

> and much of our mental life and behavioral organization functions in the service of reducing any discrepancies. When there is a discrepancy, negative affect occurs (e.g., fear, frustration) setting in motion cognitive and behavioral sequences in an attempt to move the current state of affairs closer to one's goals, desires, and preferences. If the discrepancy is reduced, then the mind can exit this mode and a feeling of well-being will follow until another discrepancy is detected, again setting this sequence in motion.
>
> Mindfulness can therefore be further conceptualized as a clinical approach to foster an alternative method for responding to one's stress and emotional distress. By becoming more aware of thoughts and feelings, relating to them in a wider, decentered field of awareness, and purposefully opening fully to one's experience, clients can abandon dysfunctional change agendas and adopt more adaptive strategies.

These therapeutic mindfulness exercises would seem a likely component of evolutionary intelligence in practice.

Equally important or perhaps even more important in the dynamics of inattentiveness are the various versions of confirmation biases and defensive fundamental attribution errors. If our behavior has led to negative or perhaps even disastrous results, we are demonstrably poor at learning from our mistakes. Our self-affirming mental posture is that it wasn't my fault, and as a result, there is nothing to be learned from the incident. It is an obvious psychological dynamic—thinking about personal failure or the partiality of partial success is uncomfortable and unpleasant, so we avoid it. One would think a phenomenon like this would be an engine stimulating systematic experimental studies in psychology, but it is a surprisingly neglected area. The field is dominated by the pop psychology of self-help publishing: "5 Ways to Turn Your Mistake into a Valuable Life Lesson," acknowledge your errors, ask tough questions, make a plan, and so on. Perhaps the reason for neglect is that we as individuals are notoriously poor at self-assessment; accordingly, these cognitive dynamics become the domain of therapeutic and clinical psychology whereby the professional "other" plays the important task of drawing attention to life lessons in environmental feedback cues. And here is yet another irony. The therapeutic community doesn't follow its own advice and collectively suffers from inattentiveness to evidence of therapeutic failures. Two British therapists analyzed the family therapy literature and found a systematic underreporting of unsuccessful therapeutic interventions, relabeling failure as "in progress" and relegating fault to patients' discontinuation of treatment. Funny thing.

But the notion of a professional therapist as the "other" who may be able to observe in the environment what the "self" cannot is important to acknowledge. The various simulated scenarios included in this chapter do have the flavor of a therapy session with the classic therapeutic expressions: Imagine yourself in the other's shoes. Why do you feel that way? Could it have come out differently?

Intelligent technologies would move the rarefied dynamics of the hundreds-of-dollars-an-hour therapists' couch to routine everyday experience for all who wish to take advantage, not just the well-to-do.

- You deducted your vacation travel as a business expense last year. It turned out to be a red flag for the Internal Revenue Service (IRS), and the notation is likely still in your records there. It might be prudent not to deduct that this year.
- Three of the five last times you went to the car dealer to window-shop, you bought a car.
- You've asked her out to lunch six times. She's been busy and unavailable each time. She's a busy gal. Have you given thought to any of the 17 other available young women in the office?
- When you end up with four remaining pills at the end of the week, you've missed four required timely doses. Let's explore a pill-taking routine tied to other reliably scheduled activities to avoid this happening again.

Isolated Thinkers—The Wisdom of Crowds

When we have to decide alone, we lack the wisdom of our family, friends, colleagues, and advisers. In a friendly trivia contest, better to have a larger team so at least one of us will know that the longest Major League Baseball game on record was 25 innings (Milwaukee Brewers playing the Chicago White Sox, May 8, 1984, with the Sox winning 7 to 6). (It is also better to have diverse team membership in terms of age, gender, and background. We'll get to the empirical evidence for that later.) Following English common law, many jurisdictions rely on a jury of 12. The judgment of the group is more reliable and judicious than a single decision maker. In ancient Greece, juries included up to 500 jurors and 1,001 jurors in capital cases.

If you have not had a chance to read it, you have nonetheless probably heard something about James Surowiecki's 2004 best-seller *The Wisdom of Crowds: Why the Many Are Smarter Than the Few and How Collective Wisdom Shapes Business, Economies, Societies and Nations*. He starts off with an impressive story from Edwardian England. The year was 1906, and the event was the West England Fat Stock and Poultry Exhibition that featured a contest. Surowiecki explains, "A fat ox had been selected and placed on display, and members of a gathering crowd were lining up to place wagers on the weight of the ox. . . . For sixpence, you could buy a stamped and numbered ticket, where you filled in your address and your estimate. The best guesses would receive prizes." He reports that about 800 fairgoers tried their luck. Although there were some farmers and butchers in the crowd who might be considered experts, most were just townspeople game to test their speculations. The average of the crowd: 1,197 pounds. The actual weight of the ox: 1,198. That is a stunning exemplar of collective wisdom. Surowiecki provides other examples but warns that the collective dynamic only works reliably when four conditions are met:

- Diversity of opinion. Each person must have some private information to bring to the group—that is, their own interpretation or their own understanding of the problem space or a related problem space.
- Independence. People hold to their own reasoning to some degree.
- Decentralization. Individuals are able to specialize and draw on their local knowledge. Someone is going to be closest to a certain aspect of the problem space, and this is what is meant by local knowledge.
- Aggregation. The means must exist to synthesize the thoughts of the team into a collective decision.

Many readers of Surowiecki's book were quick to point out that the wisdom of the crowd is basically the wisdom of the marketplace and far from a new insight into collective behavior. They point out as well the many failures of market dynamics, particularly the speculative "bubble" phenomenon of self-reinforcing but unsustainable and thus temporary popularity. So, if EI is to work as hoped, the key is a refined and prudent assessment of market signals rather than a slavish or mechanical response. If everybody shares the same wishful thinking, then the market signals will be systematically biased. That is why the sports betting coming out of Boston favors a Red Sox win. Include the bets from Yankee territory and you might have a better estimate of the outcome of a forthcoming game.

The key indicators we all recognize, of course, are market-determined prices of goods and services. But what has evolved in an increasingly interconnected and online world is "big data" reflecting virtually all of our social interactions, not just the buying and selling of a gallon of gasoline. Currently, the availability of big data resources and the computational tools and skills to analyze these immensely large and complex data sets have been the rarefied domain of big data specialists. But the future holds promise that these useful tools will ultimately migrate to your smartphone, smart glasses, and earpieces. Real-time prediction markets have expanded well beyond stocks and sporting events. It used to be that you had to call up your broker or your bookie on the phone to ask about the latest market odds and potentially place your bets. (Brokers and bookies present interesting professional similarities.) They had institutionally protected and isolated access to the market and the big data of the time. No longer.

It turns out that social scientists have been hard at work trying to understand the dynamics of collective wisdom and, of course, the equally important self-reinforcing dynamics of collective misapprehension. Productive scholarship has been growing in largely independent clusters of research under such labels as complex adaptive systems theory, behavioral economics, rational choice theory, computer-supported cooperative work, distributed cognition, prediction markets, social cognition, collective intelligence, recommendation systems, collaborative filtering, commons-based peer production, the network society, and collective behavior. The diversity of terminology and academic specializations is a bit intimidating, but the shared common ground is highly encouraging. The basic idea is that each individual's status as a node in a complex interconnected network has become more important in the digital age. Of course, there are many historical precedents—elections, crowds, markets, Nielsen ratings, public opinion surveys, hog-weight contests at agricultural fairs. But our almost constant, nearly universal electronic connectivity changes that. Presently, the "big data" created by each of our thousands of interactions each day is, as we have noted, the domain of commercial specialists. The challenge is making it equally useful, meaningful, practical, and beneficial for the rest of us. There is nothing technically preventing that. Unless commercial or political forces intervene to prevent it, there will be an app for that—no doubt many thousands of apps—and we will come to take this collective distributed cognition and its many benefits for granted.

- Your property taxes are 28 percent higher than comparable properties on your block. An appeal could lower your taxes.
- There are only three orders of prime rib left at Murray's. Order now if you still want one this evening.
- Given your past preferences and those of your peers, you're gonna love this next release by Taylor Swift.
- It is now evident that you were exposed to the new and especially contagious virus two days ago. It may be a good idea to get tested, even in the absence of symptoms.
- Ten percent of the Verizon cellular subscribers in your area have switched to AT&T because of cost and coverage issues in the last six months.

Emotional Thinkers—Sometimes Our Emotions Mislead, Sometimes Not

In everyday life, as notably in literature and popular culture, we frequently blame bad personal decisions on the emotions of the moment. Crimes of passion and temporary insanity have special status in legal proceedings. Many of us have felt an instantaneous pang of road rage even though, thankfully, most of us don't act on it. Many of us can report regretting moments of inebriated indiscretion. At such points in our lives, we could benefit from a timely voice suggesting we take a time-out to stop and think it over.

The metaphor for emotion is often one of heat, so we use the notion of a cooling-off period as one to bring the emotional and the rational back into appropriate balance. Following this metaphor, one prospect would be to use some sort of automatic fuse or circuit breaker that detects psychological overheating and responds accordingly. Such mechanisms are often built into the institutional structure of adjudications, negotiations, mediations, and markets.

The idea, however, is not simply to eliminate the emotional element entirely. Given the prominence of emotions in human cognitive dynamics, one could appropriately ask, What is the evolutionary purpose of emotion? Psychologists, it turns out, have a clear and precise answer to this question. Emotions are designed to keep our attention focused on important things, to keep us from the distractions of the unimportant. Swiss psychologist Klaus Scherer puts it a little more formally: Emotion is "an episode of massive synchronous recruitment of mental and somatic resources to adapt to and cope with a stimulus event that is subjectively appraised as being highly pertinent to needs, goals and values of the individual." In other words, our emotional systems are key to directing our attention. In my own research, my colleagues and I have found again and again that familiar and expected positive and familiar and expected negative stimuli result in a fast, routinized, and relatively unthinking System 1 response. When a stimulus is unfamiliar and knowing whether it will prove to have a positive or negative influence on our goals remains unclear, the emotional response is anxiety, and anxiety, in turn, leads to potentially beneficial information-seeking behavior.

Ask somebody what distinguishes a computer from a person and you are likely to hear that computers may be able to reason but they don't have emotions. It is the emotionless Mr. Spock stereotype from *Star Trek* again. Emotions are visceral, messy, wet, biological, and just don't translate to the clean-room logic of the silicon chip.

Tell that to Rosalind Picard of the MIT Media Lab. She has been running the Affective Computing Research Group there since the 1990s, which has spun off multiple inventions and start-up companies. She is not arguing that computers need to be more emotional than logical. She is a human–computer interface specialist. She figures that if humans are going to insist on being all emotional about their environment, then computers are going to have to figure out how to deal with that. Here's how she puts it: "If we want computers to be genuinely intelligent and to interact naturally with us, we must give computers the ability to recognize, understand, even to have and express emotions."

So her emphasis is in imbuing computers with emotional intelligence more than emotions per se. That requires the capacity to recognize subtle emotional cues in text, in speech, and in physical behavior. This is going to be a real challenge. Take physical presentations of emotion, for example. Forty distinct facial muscle movements make possible 10,000 possible combinations to create a resultant facial expression. Facial expressions, of course, are likely to be accompanied by several of 400 possible vocal inflections and perhaps several thousand hand or body gestures.

One of her spin-off companies, Affectiva, currently sells software to analyze the facial expressions of car drivers, and consumers watching advertising. It's the current cutting edge. Automotive multimodal, in-cabin sensing artificial intelligence, they claim, can identify complex and nuanced emotional and cognitive states of drivers and passengers, from their faces and voices, in order to improve road safety. Media analytic software scans faces to reveal what consumers "really feel" when they can't or won't say so themselves. A face-analyzing competitor, Emotient Inc., was bought by Apple in 2016, further signaling the recognition that these researchers are on to something that may prove to be important.

It is a little unfair to criticize programmers for their limited abilities to incorporate emotional dynamics in their human–computer interfaces. The human capacity to accurately recognize and appropriately respond to the emotional states of our conspecifics is, well, limited. Our understanding of how emotion and reason interact in the brain is elementary. Our modeling of how artificial systems could and how actual biological systems do balance emotional and logical cues will progress scientifically together and, clearly, synergistically.

Perhaps some readers are thinking—not me. If a computer has any useful advice, I'll take it straight. I don't need any smiley faces or lilting female voices with an ever-so-slight European accent. Two points can be made here.

First, the individual character of the EI interface will develop over time, as interactions are or are not particularly helpful given our lifestyles and preferences. This is just as it is for any pair of coworkers with a common project. Some of us may prefer an unshakable Spock-like voice in response to our queries. Others prefer responses given liltingly and breathlessly.

Second, we might usefully return to the evolutionary origins of emotions as that signaling system in the brain that recognizes when signals from our environment are particularly important. In the original primitive case signaling flight or fight, our adrenaline pumps, our arousal and alertness are heightened. It would seem equally appropriate for environment-scanning intelligent systems to devote additional computational power to critically important signals. Recognizing which signal is which is frustratingly important.

- Given your level of anger and frustration, it may be a good time for a break.
- Despite the many smiles in the room, I'm sensing an unusually high level of tension and fear.
- The class is getting sleepy, Professor. Time to change topics and reenergize learning activities.
- Your heart rate and perspiration is spiking. Think about why that may be the case.
- You are nervous. Try using the usual breath control techniques to calm down.

So what makes compensating for our hardwired cognitive biases so hard? Answering this question is critically important to the actual prospect of successful evolutionary intelligence. The short answer is that, unfortunately, we have become pretty good at ignoring good advice when it is offered. We confidently overestimate our understanding of our environment. We proudly ignore the sage advice of our parents as we complain that they just "don't understand." We succumb to the pull of transitory emotions or social pressures. We're in a hurry. We may even tacitly recognize the wisdom of some practical advice, but we still don't like it. We opt to ignore it.

The current stage of applied artificial intelligence might be characterized as good at a few things and pretty lousy at most of the others. If we ask Siri or Alexa for advice today, the results are likely to be hit or miss. Some of the missed cues are hilariously off-kilter. We have come to understand that our smart speakers are OK at understanding our speech and very good at looking stuff up—time of day, weather, sports scores, meatloaf recipes. We are unlikely to ask for advice on matters of romance, business investment, or career strategies.

In the next few decades, our evolved biological intelligence will remain for all practical purposes unchanged. Evolutionary selectivity takes hundreds and thousands of generations. So we will continue to make mistakes in our life's decisions at pretty much the same rate as before.

What will change in the next few decades, however, is the logical and communicative capacities of artificial intelligence. The progress is likely to be exponential—by all measures, a game-changer.

So will we actually change our behavior to take this new potential of evolutionary intelligence into account?

Not all of us. Not right away. Not in every instance. But in time, largely yes.

The wheel made us more mobile. Machine power made us stronger. Telecommunication gave us the capacity to communicate over great distances. We've learned to take the car, to use the mixer or power drill, to send a text message rather than yell out the window. We do it naturally and unthinkingly. We have learned that intelligent GPS-based services like Waze are pretty good in providing directional advice. Even veteran cabbies who know the traffic well find it useful to have their cell phone on the dash monitoring traffic. We see others taking fruitful advantage. We come gradually to appreciate the power of these technologies. Applied, everyday artificial intelligence has spectacular potential. And given the challenges of our fragile environment and global tensions, it may arrive just in time.

4 Here Be Dragons

Roman and medieval cartographers developed the tradition of drawing sea monsters and lion-like creatures to designate the unknown dangers of the uncharted lands and oceans at the edges of their maps. Mystery implies danger. Just to make it clear, some mapmakers even wrote out the text in the margins—*hic sunt dracones*, here be dragons. There are many unknowns about how the phenomenon of evolutionary intelligence (EI) will ultimately become a routine part of our lives. And just as it was with each preceding generation of technology, there will be individuals with malevolent or criminal intent who will try to harness the power of these technologies to do evil things. We would be remiss not to look more closely at these potential fault lines in the future.

We have characterized EI as our digital interface with the world around us. Like the clothes we wear, the car we drive, our social media avatar, and our personal website, the way we design and use helpful technologies says something about us and represents our mediated interface with our environment. We wouldn't want to rely on a dysfunctional or counterproductive interface intentionally. We want to look our best and be successful. We want reliable and classy transportation if we can afford it. What could go wrong?

You saw this coming? Yes, a lot could go wrong. The classic answer to what could go wrong is that the artificial intelligence gets so intelligent it "decides" to serve its own interests rather than those for which it was designed. The killer robot. Design a robot with lethal capacity and it decides to kill its creator. Or what may be the seminal narrative of a computer that has decided to ignore human commands and take over on its own—HAL 9000 in the 1968 space adventure movie written and directed by Hollywood icon Stanley Kubrick, *2001: A Space Odyssey*. It takes place on a space station inhabited by surviving astronaut Dr. Dave Bowman, played by Keir Dullea, and the blinking red light that represents the central computer HAL that has taken over control of the station, voiced with a mechanical calm by Canadian actor Douglas Rain. Dullea's character is trapped outside returning from a spacewalk in which his partner astronaut had been killed.

Dave: Hello, HAL. Do you read me, HAL?

HAL: Affirmative, Dave. I read you.

Dave: Open the pod bay doors, HAL.

HAL: I'm sorry, Dave. I'm afraid I can't do that.

Dave: What's the problem?

HAL: I think you know what the problem is just as well as I do.

Dave: What are you talking about, HAL?

HAL: This mission is too important for me to allow you to jeopardize it.

Dave: I don't know what you're talking about, HAL.

HAL: I know that you and Frank were planning to disconnect me, and I'm afraid that's something I cannot allow to happen.

Dave: [feigning ignorance] Where the hell did you get that idea, HAL?

HAL: Dave, although you took very thorough precautions in the pod against my hearing you, I could see your lips move.

Dave: Alright, HAL. I'll go in through the emergency airlock.

HAL: Without your space helmet, Dave? You're going to find that rather difficult.

Dave: HAL, I won't argue with you anymore! Open the doors!

HAL: Dave, this conversation can serve no purpose anymore. Goodbye.

Dave manages to reenter the space station and ultimately physically disconnects HAL's higher-level functions as HAL pleads for forgiveness and reverts to its early programming, singing a child's song it first learned many years ago back on earth. It is high-powered drama and in many ways the pivotal core of the film. The issue of diverging purposes between human and machine, of course, are more complex.

In my view, there are four types of danger, each a broad domain of dysfunction that could apply as well to any sociotechnical transition in history.

First, the early-stage EI interface could make an innocent mistake —more or less a miscellaneous error of various sorts. It could misunderstand the situation, misinterpret social or economic signals, awkwardly reveal personal information to others, or basically give very bad advice. As these technologies are trialed in early versions, there are likely to be many instances of this occurring. Some will be humorous, such as when Siri makes a faux pas of some sort. Some technical errors may lead to serious negative consequences—perhaps an economic loss or personal injury. There have already been fatalities attributed to self-driving cars. Presumably, we humans will be highly motivated to recognize and correct misinterpretations and bad advice because ongoing feedback will continuously improve performance. As I see it, this is primarily about tuning the technology to function as we intend it to. I won't dwell on this first case.

Second, EI processes could be hijacked by others with malign intent. This situation is serious, fundamental, and important. It could be a particular individual with whom you are negotiating. It could be unknown third parties with nefarious interests of their own. As with any social system like paper currencies, lotteries, bank loans, credit cards, eBay-like trading systems, and stock exchanges, there will be fraud. We will address this issue in the pages ahead.

Third, artificially intelligent EI processes could function to serve their own interests rather than ours. This, of course, draws on the classic fictional dystopian scenario of the war of humans versus machines—the HAL computer, killer robots, the Matrix, replicants, the Terminator. Another serious issue worth serious attention.

Fourth, systemic developments could have unintended negative consequences, not necessarily self-interested or criminal manipulations, just unanticipated bad outcomes. This is also an important consideration but particularly difficult to address. We move, in the vocabulary of Donald Rumsfeld, from known unknowns to unknown unknowns. As it is said, in each case, here be dragons. We can draw on historical lessons and design early-warning measures to try to minimize risk.

OK, here we are in the dragons chapter. This is where we review all the nasty things technology has wrought and has yet to inflict. Don't get me wrong. These are serious questions. But I'm a bit wary of technology bashing. I'm with historian Mel Kranzberg, whose first law of technology is:

> *Technology is neither good nor bad; nor is it neutral.*

Kranzberg was a serious historian of technology, and I take his aphorism to mean that technologies in themselves don't intrinsically harbor good or evil but that in their interaction with individuals and institutions, they may well tip the balance, especially when they first arrive. Best we pay close attention. So the takeaway is that the balance of positive and negative pivots on how technologies are implemented. And, in turn, given my argument about inevitability, we may need to get this next one right.

The problem, however, for those of us in this field is that technology bashing is convenient, narratively engaging, reassuring in that it shifts the blame, and resonates with conservative sensibilities that favor older traditions. I have a favorite moral tale about technology bashing that I sometimes recount from my study of mass communication effects. The story is told by colleagues Ellen Wartella and Byron Reeves, who reviewed the recent history of research of media effects on children. The basic idea is that as soon as a new medium comes along, it is instantly responsible for all that is wrong with our kids. Movies caused juvenile delinquency. Radio frightened young folks. Comic books interfered with kids' reading skills. Television did all of the above. (Video games were just becoming popular as Wartella and Reeves drafted their study, so various demonstrations of its evil effects were still in the planning stages.) They quote a particularly breathless critic of radio from the 1930s:

> The popularity of this new pastime [radio] among children has increased rapidly. This new invader of the privacy of the home has brought many a disturbing influence in its wake. Parents have become aware of a puzzling change in the behavior of their children. They are bewildered by a host of new problems, and find themselves unprepared, frightened, resentful, helpless.

This chapter has a unifying thematic, a strategic idea for dealing with the diversity of dragons in these future domains not yet fully explored. The strategic idea is transparency. As decisional algorithms evolve, they need to be accessible so that biases and errors can be identified and, if the social will is strong enough, made right. Individuals and institutions that might benefit from biases, naturally enough, prefer obfuscation and mysticism—wizards at work, so, Dorothy, pay no attention to that man behind the curtain.

The industrial West has a strong tradition of private commercial control of industrial systems and processes that operate through a web of copyrights, patents, and trademarks. The tradition encourages market rivalry of competing internally coherent, closed-source proprietary systems. More recently, however, the software industry has spawned a vibrant alternative model of open-source software and systems. The Internet's core protocol, the TCP/IP software platform, has always been open source, available to all without cost and open to inspection, improvement, and correction. Apple and Microsoft tend to favor proprietary models, Google a mix of open and closed. So it is not clear that proponents of open source will dominate the future of artificial intelligence (AI) algorithms. Even if that is the case, or partially the case, proprietary systems do not prevent transparency because such systems can be tested on a "black box" basis with clear-cut inputs and outputs to test for potential biases and misrepresentations. Let's take a closer look at the likely dragons in our future.

Corruption—Artificially Intelligent Crime

We all have a collection of fragmentary scenes from old movies watched long ago still prominently stored in our heads. For me, there is a fragment from a 1930s black-and-white crime drama whose provenance and actual title were long ago forgotten. But, in this fondly remembered scene, there are two career criminals just back on the street from Sing Sing, ready to get back to their criminal ways. Technology marched on while the two were up the river. One warns the other: "Hey, you gotta be real careful. The cops now have these two-way radios in their cars." Cinematic cops and robbers with technology evolution playing its role. The quintessential 1930s cop Dick Tracy even had a wristband two-way communicator. The cartoonist dutifully denoted the wristband with an arrow and text carefully printed out for the uninitiated: "2-way wrist TV." Crime marches on as well, and criminal strategies adjust. If two-way radios gave the cops a relative advantage, it was likely fleeting. Now we all have two-way radios on our wrists or at our hips. It is not clear that levels of crime have changed in any meaningful way as a result and as the nature of crimes themselves in the digital era evolve. Can we make some sense of whether EI might somehow tip the scales in the ongoing battle between the good guys and the bad guys?

The short answer is yes, and not in the right direction. To the extent that the complexity of digital decision algorithms makes them less accessible to those who use them, then they are increasingly susceptible to hidden manipulation by third parties. But that is as true of the current digital era as of evolving artificial intelligence. We don't inspect the code of our tax software or the code behind our broker's statement or our cable bill for errors. We may respond to anomalies, and there are systems for audit and oversight. So, as expected, the answer is that because the algorithmic complexity cannot be avoided, independent audit and output testing are more important than ever. Are the banks redlining? Did the electoral commission report the vote accurately? Does Google surreptitiously manipulate search results to heighten the visibility of its own products and services? Such questions require technically sophisticated independent assessment. Transparency.

Another potential issue could be characterized as gaming the system. If you know the algorithmic rules, you can tweak your responses to your advantage. Take the automated grading algorithms for standardized SAT-type essay exams, for example. The algorithms use measures of sentence length, spelling accuracy, and word-choice diversity to compute essay grades. Although it is technically possible to invent multisyllabic nonsense gobbledygook paragraphs that receive high scores, the systems, when double-checked by human graders, are usually right on target. Generative AI technologies like ChatGPT are likely to raise the stakes in the gaming of essay writing.

Copy and paste makes student plagiarism easier to undertake and, of course, equally easy to detect. Systems and system users each use increasingly sophisticated algorithms to initiate and, in turn, to identify gaming. Gaming the system is not new. It will surely continue.

Temporary advantages will be compensated for in what could be characterized as a classic technological arms race. There is no reason to believe the playing field has become permanently tilted.

Misrepresentation and Truth Decay

Computers have gotten increasingly good at blurring the distinction between the real and the unreal. Computer-generated graphics dominate the motion picture industry. Photoshopped images are de rigueur in advertising. Computers auto-tune the vocalist in recorded music. Call the help desk and speak to an empathic but robotic voice. And among the most troubling of very sophisticated unreality is the deep fake.

All it takes is about 20 minutes of aggregated raw video of a speaker to capture all of the possible phonemes and most of the possible facial and lip movements. Slice the video into fragments and stich it back together into an apparently seamless collage and voilà. You can make the speaker say virtually anything you want. When done well, it is almost impossible for the untrained eye to distinguish the re-created deep fake from the real. Convincing exemplars of world leaders, politicians, and celebrities enthusiastically uttering gibberish or making outrageous public pronouncements are routinely circulated on the web. Predictably most of the amateur deep fake creations involve a simple face swap by pasting celebrity or acquaintance faces on pornographic images or videos. DeepTrace Technologies recently identified 15,000 deep fake videos online, 96 percent of which were in fact explicit sex videos.

Faked audio or video can represent a serious criminal enterprise. Criminals used AI-based software to impersonate a chief executive's voice and demand a fraudulent transfer of about a quarter million dollars in 2019. The head of an English energy company thought he was speaking on the phone with the chief executive of the firm's

German parent company, recognizing the voice and the slight German accent. The voice asked him to urgently send the funds to a Hungarian supplier within the hour. He did. The crooks got away with the loot. Fortunately, the victimized company was fully covered by insurance.

Equally troublesome may be the shallow fake. This is a video or audio edited just slightly to omit words or make the speaker sound drunk or disoriented. In one prominent victim was then Speaker of the U.S. House of Representatives Nancy Pelosi. The videotape was amateurishly edited to slow down her speech at various intervals so that she sounded drunk and confused. The doctored video was both widely circulated and widely discredited.

The predominant terminology, perhaps now overused, is fake news. An independent and trustworthy source of news about public affairs is a prerequisite for democracy. Thomas Jefferson famously summarized his views on the importance of a reliable free press: if left to decide on "a government without newspapers or newspapers without a government, I should not hesitate a moment to prefer the latter."

The prominence of falsehoods, conspiracy theories, distractions, and exaggerations in politics and public life is also not new. It is as old as public life itself. What may be new is a shifting balance in the public's enthusiasm to embrace them.

Robert Putnam captured public attention in 2000 with his "bowling alone" thesis, and he found a convenient foil deserving blame—television. Americans, he argued, spend too much time at home alone watching TV and no longer developing social skills and social capital by engaging in community activities like unions, political clubs, and bowling leagues. Since viewing had been trending up and bowling and union membership trending down, it is easy to impute an apparent causal connection. Although there is no way to prove television did or did not contribute in some small way, it would seem apparent that a much broader social trend is underway that has led to increasing cynicism, distrust, conspiratorial imaginativeness, and dramatic political polarization. The current culprit is the Internet rather than television. But the trend started before either technology became dominant. In the 1950s and 1960s in the U.S., as postwar enthusiasm began to wane, public trust in the government and in the media began to decline. It is, in part, likely a result of the culture wars of the 1960s, Vietnam, and Watergate, working up to the ultimate polarization of the Trump era, which is dominated by exchanged critiques from the left and the right of fake news, alternative facts, and what could appropriately be characterized as truth decay.

It may or may not have been Mark Twain who first said that "a lie can travel halfway around the world while the truth is still putting on its boots." (It does sound like Twain.) But the point may have been more characteristic of his era than ours. Truth and falsity are both instantaneous. Virality is simply not a function of truthfulness but rather the narrative resonance of a fact or story with segments of the public that are amused or intrigued. Is there reason to believe evolutionary intelligence will interact with these trends?

There is. And to derive an answer, we turn to Nicholas Negroponte's now-famous technological notion of the "Daily Me." The original idea was straightforward and innocent, but it has become dramatized. The idea is simply an electronic newspaper attuned to your interests. Negroponte proposed:

> What if a newspaper company were willing to put its entire staff at your beck and call for one edition? It would mix headline news with "less important" stories relating to acquaintances, people you will see tomorrow, and places you are about to go to or have just come from. It would report on companies you know. . . . Call it *The Daily Me.*

Subsequent analysts characterized this as a dangerously polarizing filter bubble and echo chamber that would reinforce polarization and relative ignorance of the views of others. Perhaps the critics did not manage to read Negroponte's following paragraph:

> On Sunday afternoon, however, we may wish to experience the news with much more serendipity, learning about things we never knew we were interested in. . . . This is The Daily Us. The last thing you want on a rainy Sunday afternoon is a high-strung interface agent trying to remove the seemingly irrelevant material. These are not two distinct states of being, black and white. We tend to move between them, and, depending on time available, time of day, and our mood, we will want lesser or greater degrees of personalization. Imagine a computer display of news stories with a knob that, like a volume control, allows you to crank personalization up or down. You could have many of these controls, including a slider that moves both literally and politically from left to right to modify stories about public affairs.

His choice of words is interesting. He is not imagining an ideological human editor insulating you from possible provocation but an intelligent agent under your control, as he puts it, a very sophisticated volume control, at your personal disposal. That's evolutionary intelligence.

Is there reason to believe that, over time, factual evidence will accumulate so that actual facts will be sustained and alternative facts dismissed? The answer is yes, but if and only if those of us who have the volume control in hand insist. The good news is EI makes it easier to implement.

Intelligent Personal Privacy

The Israeli historian and intellectual celebrity Yuval Noah Harari is fond of noting that in the modern age, the most important thing each of us has is our personal information, and we thoughtlessly give it away in order to watch funny cat videos.

Let's take a closer look at his argument. Harari surveys the dominant corporate titans of our age and observes correctly that the economic power of the likes of Google and Amazon is largely derived from their capacity to monetize what they know about us. Google's annual advertising revenues are over $130 billion because advertisers would much rather draw attention to their brand of, say, athletic shoes precisely when an individual is surfing the web for sports gear rather than when a random viewer is watching a random television program. Amazon, in turn, is very astute at selling you a new Jack Reacher novel because it knows you've already bought a bunch of them. Harari's cat-video meme is his way of signaling that it is casual amusement and convenience for which we trade our valuable personal information. He goes on to note that we have thousands of years of experience in regulating the ownership of land and personal goods but none at all in the ownership of personal data. Then, uncharacteristically, he throws up his hands in puzzlement and defeat and wonders if Mark Zuckerberg and his 2 billion friends could figure this out. Harari proposes personal control of personal data. Sounds like a good idea.

Three things are necessary for this to work. First, there needs to be general recognition that Harari (along with just about everybody in Silicon Valley) is right that personal information is a valuable commodity. Second, there needs to be some sort of regulatory regime in place to sustain the marketplace for this commodity as there currently is for physical products, services, intellectual property, works of art, and even likenesses. (Celebrities have, after all, legally and routinely monetized their likenesses and implied endorsements in return for millions of dollars.) These extant regulatory regimes provide abundant legal remedies to prevent theft, unlicensed use, plagiarism, and other violations. And third, there needs to be some sort of bookkeeping system to manage the data, usage, transfer of rights, expiration of rights, and related details. This is precisely the sort of enterprise that evolutionary intelligence should be good at.

Let's take the example that individual X is a connoisseur of fine chocolates and both inclined to and financially positioned to purchase ample amounts of same. Traditionally, vendors in this marketplace would rely on advertising to reach subscribers of *Cacao, Chocolate Affairs,* or *Chocolate Connoisseur Magazine* with their appropriately high-end advertising rate cards for a highly targeted clientele. If Ms. X's personal predilections and purchase history are a valuable commodity, why shouldn't that value accrue to Ms. X herself? Cut out the intermediary. The reason it hasn't happened yet, of course, is that the recognition, the regulation, and the technical mechanism have not yet evolved. But given the aggregated financial values involved, it is a good bet that they will.

How would this work? Like buying an ad in *Cacao,* it is a negotiation between a product vendor and an intermediary on behalf of a potential customer. Each of us is seen by the marketplace as a dynamic aggregation of various purchase probabilities. Ms. X has a 50 percent chance of a chocolate purchase, a 20 percent chance of an athletic shoe purchase, and a 0.01 percent chance of a ride-on lawn mower purchase because she lives in a townhouse with no yard. The typical marketing cost for major purchases is substantial. For automobiles, it is $700 to $1,000 per purchased car. The ingredients in a typical bottle of perfume cost about $1.50. The bottle sells for $150.00. The rest is packaging and promotion and advertising of various types. If your electronic interface—that is, your evolutionary intelligence—were to announce that you were in the market for $1,000 of clothes and cosmetics or a new car for $75,000, that is valuable and monetizable information, monetizable by you.

Wait. What? This section began under a heading of "personal privacy." What is all this dissertating about just the opposite—systematically making personal information available to other entities? Actually, I believe that is entirely the point. Intelligent personal privacy makes it routinely convenient for each of us to take control of our personal information. If you don't want anybody to know of your weakness for fine chocolates, so be it. When you purchase fine chocolates or download a recipe for chocolate truffle ganache, you keep your identity contractually private and happily forfeit the financial benefit or discount. That is, of course, possible today—just shop at CVS and opt out of using the CVS discount card that matches your purchases with your identity. Pay with cash. Further, one can use a virtual private network and specialized browsers and search engines that use sophisticated techniques to mask your identity.

Personal privacy in the twenty-first century is a curious domain of strongly held beliefs. Many express deep concern that distant corporate and government databases know too much about them. Many find it unnerving that after they have spent a little time online looking over toaster ovens, pictures of shiny toasting devices start appearing in ads all over their online sports news. Your search behavior is monetizable, and the company that provided your free browser software is doing just that. You can opt out and insulate your identity in various ways, but it takes effort and some technical sophistication. Search software companies have no incentive to make it easy for you.

Folks who write computer code and build online systems as a community share a unique culture, one that tends to take individualism and privacy rights seriously. So, sure enough, there have been numerous, largely spontaneous efforts to address the privacy problem. Among them was P3P, the Platform for Privacy Preferences Project, which built a protocol to permit websites to declare their intended use of the information they collect from web browsers. It was designed by web pioneer Tim Berners-Lee and the World Wide Web Consortium at MIT in 2002. It was implemented briefly by Microsoft's Explorer browser but few others and then quietly dropped out of sight entirely. Critics claimed it was too complex and difficult for the average person to understand and implement. Further, there was no way to enforce any website's implied promise not to make use of personal data. No records are kept, and there are no marketplace or legal sanctions for violations.

In 2009, a group of privacy-oriented activists proposed an optional header that could be routinely included in a web browser and that declared explicitly to each website addressed: "Do Not Track." A variety of popular browsers took up the option briefly, but it too was ultimately dropped and abandoned entirely by 2019. Critics complained that, like P3P, this approach was ineffective and lacked external enforcement. Online advertisers insisted that if P3P was installed as a default option, it would violate their contractual agreements with websites. At its peak, only about one in ten users opted for the Do Not Track (DNT) option, and only a handful of websites honored the request when it was sent. And as one observer put it, as if directly from the Department of Irony, Google's Chrome browser offered users the DNT option, but Google itself didn't honor the request.

So what would seem to be like a slam-dunk technical option turns out to be an airball that's not close to the rim or even the backboard. The reason for this, in my view, is that the three perquisites noted above have not yet been met: (1) public recognition of the value of personal information, (2) some form of regulatory or market enforcement, and (3) a convenient, reliable, unobtrusive bookkeeping system.

My scenario for the future of intelligent personal privacy is a tipping-point phenomenon. Take note of two common elements of the failed efforts described above: the hassle factor and the economic incentive factor. Keeping track of whether websites are or are not abiding by privacy agreements is getting easier. It may involve some form of the blockchain public ledger technology, with multiple copies of audit chain information stored publicly and available for review. And it will likely engage the notion of "smart contracts" through which the software actively monitors, over the network, contractual elements to ensure they have been completed as required. Monitoring compliance will not work if it requires the casual web user to email or make a phone call to complain. Most casual users wouldn't follow up anyway because the privacy question is only of some modest concern to them. The tipping-point phenomenon comes when one casual user recognizes that another casual user is actually profiting financially from taking control of their personal information. That positions the financial interests of millions of users (so far a diffuse and partially articulated interest) against the focused and fully articulated interests of online advertising and targeting mediators like DoubleClick, now owned by Google. When the public interest is manifest and the bookkeeping technology is hassle-free, the transition to personally controlled personal information will happen, with or without government intervention. Smart contracts are able to work largely within the current and well-established tradition of contract law.

Privacy advocates and activists have been working at a distinct disadvantage because large majorities of the public simply do not care much about these issues. If Google knows many details about my love of chocolate and of Vampire Weekend music, so be it. To put things in perspective, it may be useful to think historically for a moment about public expectations. Until the end of the nineteenth century, most of us lived not in anonymous urban complexes but in small, largely rural communities where everybody knew everybody else. And, with relatively few exceptions, everybody knew everybody else's business. In 1840, roughly one in ten Americans lived in urban areas. Now more than eight in ten do. Sociologists, historians, novelists, and journalists carefully monitored the transition from the small town, where everybody knew your name, the names of everybody else in your family, and probably how well your farm did last year, to the anomic and anonymous crowds of the industrial city. The prominent concern during this transition was not privacy but lack of community connection—the mass society, the lonely crowd, the breakdown of social norms. Although we may no longer be deeply concerned about being lonely in the crowd, perhaps we should not be surprised that not everybody is deeply concerned about being anonymous in the crowd.

Cognitive Atrophy from Dependence on Intelligent Technologies

Perhaps the most seminal exclamation of concern about mental decline from dependence on exterior resources comes from the celebrated Phaedrus dialogue. Socrates explains that a supplicant proposes to the King of Egypt that all Egyptians be taught reading and writing so that they will be wise. King Thamus responds:

> You are the father of writing, your affection for it has made you describe its effects as the opposite of what they really are. In fact, it will introduce forgetfulness into the soul of those who learn it: they will not practice using their memory because they will put their trust in writing, which is external and depends on signs that belong to others, instead of trying to remember from the inside, completely on their own.

It is hard to imagine a potentially persuasive argument that literacy is bad for you, but there it is. To depend on external documents depletes your capacity to memorize. I expect that versions of this story have been retold, in turn, as the abacus replaces the fingers, the calculator replaces the slide rule, and the smartphone replaces all sorts of things.

In his cover piece for *The Atlantic* in 2008, journalist Nicholas Carr wrote:

> Over the past few years I've had an uncomfortable sense that someone, or something, has been tinkering with my brain, remapping the neural circuitry, reprogramming the memory. My mind isn't going—so far as I can tell—but it's changing. I'm not thinking the way I used to think. . . . "The perfect recall of silicon memory," Wired's Clive Thompson has written, "can be an enormous boon to thinking." But that boon comes at a price. As

> the media theorist Marshall McLuhan pointed out in the 1960s, media are not just passive channels of information. They supply the stuff of thought, but they also shape the process of thought. And what the Net seems to be doing is chipping away my capacity for concentration and contemplation.

The piece is titled: "Is Google Making Us Stupid? What the Internet Is Doing to Our Brains." He followed up that article with a full-length book a few years later. Clearly, there is a sizable market for these speculations. Many of us find it convenient and somehow satisfying to blame Google for whatever ails our cognitive capacities. In 2010, *The New York Times* initiated a series called "Your Brain on Computers," which drew your distracted attention with the following headlines:

Digital Devices Deprive Brain of Needed Downtime

Growing Up Digital, Wired for Distraction

An Ugly Toll of Technology: Impatience and Forgetfulness

More Americans Sense a Downside to an Always Plugged-In Existence

Hooked on Gadgets, and Paying a Mental Price

While at your local bookstore, you may see the following book titles beckoning you:

The Shallows: What the Internet Is Doing to Our Brains

iBrain: Surviving the Technological Alteration of the Modern Mind

Data Smog: Surviving the Information Glut

Distracted: The Erosion of Attention and the Coming Dark Age

The theme is consistent and coherent. Technologies that assist the brain are likely to weaken it. It is demonstratively true that we are less likely to recall the phone numbers of friends and families now that they are conveniently stored just a screen tap away on our smartphones. But perhaps if we are freed from memorizing otherwise meaningless phone numbers, we may be able to put our short-term and long-term memories to work on more important challenges.

King Thamus is correct that our raw capacity to memorize may retreat a bit as we become literate adults. What the good king fails to note is that without literacy in the first place, there may not be much we can find worth memorizing. I conclude that the atrophy issue is understandably engaging, perhaps even a bit romantic about olden times, but it represents for the most part a distraction from more significant concerns.

Inequities

The persistence of dramatic inequities between the rich and the poor remains one of the most troubling features of the modern world. More troubling still, despite some earnest efforts by institutions of public education and social welfare, the inequities remain largely socially, culturally, and economically reproduced from one generation to the next. The question has drawn intense concern and stimulated an immense and complex research literature. One hesitates to step into the fray . . . but we should ask: Is there some small chance, given convergent institutional support, that evolutionary intelligence might help to reduce such inequities and their social reproduction? If so, how? By what possible mechanism? Or, should we be concerned that these technical developments may widen these gaps?

Critics of technology have been very active on this issue. They are pessimistic. They assert that these evolving intelligent technologies will exacerbate inequality for three reasons: (1) Expensive new technologies will limit access for the less-well-to-do, reinforcing a digital divide; (2) marginal communities may lack the skills, training, and digital savvy to fully take advantage of evolving technologies; and (3) automation-related job losses will disproportionally affect the lower strata of occupations. Let's review these three questions in turn.

Expense is the first issue. When technologies are brand-new, they tend to be rare, expensive, and noteworthy. The relatively well-to-do are likely to be first to take up the technologies, in part to flaunt their wealth. Think of the first family on the block with color television or the first college kid with an Apple watch. But the novelty and expense will decline over time. In the case of television, for example, at one point in the late twentieth century, a larger percentage of U.S. homes had television than indoor plumbing. In the early days of the Internet, the phrase "digital divide" was used to draw attention to differential levels of network access associated with variations in socioeconomic status and geography. As those inequities declined, attention was drawn to differential broadband access speeds and, importantly, internationally, different levels of development of information infrastructure across nations. The good news is that, like the diffusion of television, Internet access has increasingly become nearly universal in the developed world.

A key element of this process has been the migration of access from a desktop or laptop computer to a smartphone. One review summarizes the pattern:

> Mobile phones have diffused across the globe faster than any other communication technology in history. Subscriptions for mobile phone service now exceed the global population, due in part to multiple-subscribing individuals. It is projected that by 2022, 90% of all mobile subscriptions will be for Internet-enabled "smartphones," which are already in the majority. In the United States, more time is spent on digital activity on smartphones than on computers. Mobile phones are credited with expediting Internet access for the global poor.

The expansion of Wi-Fi hot spots and high-speed 5G connectivity will reinforce the growing dominance of mobile access to networks. We can view this as the process we have characterized as digital conjunction: Computational intelligence moves from a remote room-sized computer to the desktop, then to the laptop and the smartphone and then to smart glasses. At each stage, the computational capacity becomes smaller, relatively less expensive, easier to use, more network-dependent, more frequently used, and physically closer to the user.

The second issue is skills. This is sometimes referred to as the second-order digital divide. Early research on Internet access pointed out that if we are interested in how these technologies may actually affect life chances, then we need to study more than just a binary notion of access to the web or its absence. In a series of seminal studies, sociologist Eszter Hargittai demonstrated that the ability to find information online varied with online experience, age, and social status. Since that work in the early 2000s, web search resources have improved, modal levels of experience with online search have increased dramatically, and special-purpose apps for mobile platforms have diminished the importance of potential skill differentials. Hargittai points out that if higher socioeconomic status is associated with higher levels of employment-related and political information online, it may be inappropriate to characterize that as a "technological effect." Drawing on Pierre Bourdieu's concept of social capital as a set of practically empowering skills for advancement in modern societies, Hargittai notes: "Traditionally, research in this domain has tended to treat social capital as a result of information and communication technology (ICT) uses, overlooking the possibility that level of social capital may be an important predictor of how people use ICTs in the first place." She draws particular attention to maintaining professionally relevant social networks online.

Research on younger mobile phone users in minority communities has found that the primary use of smartphones for these younger individuals is expressive—as in social media and entertainment, as in games rather than instrumental uses such as seeking information, education, and employment. It is an old and familiar story in the study of "communication effects." The technology does not mechanically determine how it might be used; rather, it facilitates the uses of the existing cultural environment. So, if EI resources are to contribute to reducing recurring patterns of inequity, it would have to be part of a broader cultural and economic effort.

The third issue concerns employment patterns. A great deal of attention has been paid to the impact of computerized automation on employment patterns. The oft-cited and stereotyped contextual question is: Why hire a person if a robot can do the job better, for longer hours, without complaint, without organizing unions, and without the need for retirement or health benefits? Economists who have studied the issue point out that it is not a question of simple robot-for-person substitution but a much more complex and also age-old pattern of evolving job descriptions. As Erik Brynjolfsson and colleagues point out:

> A shift is needed in the debate about the effects of AI on work: away from the common focus on full automation of many jobs and pervasive occupational replacement toward the redesign of jobs and reengineering of business processes. . . . An occupation can be viewed as a bundle of tasks, some of which offer better applications for technology than others.

They proceed to ask if machine learning may end up creating more jobs because it influences entire business processes, not just singular job descriptions. It is refreshing to move away from the dichotomous robot-or-not thinking to explore how evolutionary intelligence might enhance the capacity for retraining and teaching, as it might be put, old workers new tricks.

So, in our discussion of technology and inequality in these three instances, we have two competing theories with opposite conclusions. The Matthew Effect Theory (the tendency for early advantages to multiply over time, as in the rich get richer, from the Gospel of Matthew) is that higher social strata will have the skills and resources to better take advantage of new technologies, thus reinforcing patterns of inequality. In contrast, there is the Leap Frog Theory, which points out that those lacking informational and social capital stand to benefit most from intelligence-augmenting technologies. As the analogy goes, Einstein wouldn't benefit much from these technologies because he already had such a thoughtful grasp of how to evaluate alternatives.

My best guess is that both mechanisms will be in play more or less simultaneously at the outset. My further estimation is that as these intelligent technologies become routinized, the Matthew Effect will lessen over time and the Leap Frog dynamic will become increasingly relevant, leading to a net positive result. I could be wrong. Much will depend on how various subcultures come to define these resources as relevant or fashionable.

Self-Interested Technology—A Concern about Revengeful, Potentially Lethal Artificial Intelligence

As legend has it, while vacationing with friends in a castle on Lake Geneva, a teenage Mary Shelley was frustrated that she could not respond to Lord Byron's late-evening challenge to each person in the group to conjure up a truly terrifying horror story. Then a vision came to her several days later in a dream. That vision was ultimately published in 1818 as *Frankenstein; or, The Modern Prometheus,* and an iconic and extremely influential cultural allegory was born—the monster of Dr. Frankenstein. The notion of a powerful but ill-conceived creation beyond the control of its inventor captivated the wider public. It is reported that more than 500 editions of the novel are currently in print, that over 50,000 copies of the novel and its variants sell annually in the United States, and that the work has been translated into over 30 languages. Scientific hubris. Invention run amok. A resonant cultural theme.

A century and a half later, at the dawn of the computer era, scientists and philosophers would return to her thesis with their concerns about the well-intentioned creation of intelligent agents whose capacities would exceed our own and who might maliciously or perhaps even just whimsically destroy their creators.

Alan Turing said the following about independent machine intelligence in a speech in London in 1947:

> It has been said that computing machines can only carry out the processes that they are instructed to do. . . . Let us suppose we have set up a machine with certain initial instruction tables, so constructed that these tables might on occasion, if good reason arose, modify those tables. One can imagine that after the machine had been operating for some time, the instructions would have altered out of all recognition. . . . In such a case one would have to admit that the progress of the machine had not been foreseen when its original instructions were put in. It would be like a pupil who had learnt much from his master, but had added much more by his own work. When this happens I feel that one is obliged to regard the machine as showing intelligence.

Turing's friend and colleague Irving Good in 1965 said this about a machine intelligence explosion:

> Let an ultraintelligent machine be defined as a machine that can far surpass all the intellectual activities of any man however clever. Since the design of machines is one of these intellectual activities, an ultraintelligent machine could design even better machines; there would then unquestionably be an "intelligence explosion," and the intelligence of man would be left far behind. . . . Thus the first ultraintelligent machine is the last invention that man need ever make, provided that the machine is docile enough to tell us how to keep it under control.

As one of the prominent leaders of the concerned community, British philosopher Nick Bostrom expanded on his oft-repeated paper clip scenario from 2003:

> Suppose we have an AI whose only goal is to make as many paper clips as possible. The AI will realize quickly that it would be much

> better if there were no humans because humans might decide to switch it off. Because if humans do so, there would be fewer paper clips. Also, human bodies contain a lot of atoms that could be made into paper clips. The future that the AI would be trying to gear towards would be one in which there were a lot of paper clips but no humans.

Stephen Hawking, writing with colleagues in 2014, noted:

> If a superior alien civilisation sent us a message saying, "We'll arrive in a few decades," would we just reply, "OK, call us when you get here—we'll leave the lights on"? Probably not—but this is more or less what is happening with AI. Although we are facing potentially the best or worst thing to happen to humanity in history, little serious research is devoted to these issues.

Bostrom's colleague, MIT physicist Max Tegmark, at the outset of *Life 3.0*, tells his own little horror story, paraphrased here:

> A clandestine corporate project team tests its superintelligent machine they've dubbed "Prometheus" in secret over a long holiday weekend just to see what it can do. The draft AI program they had created redrafts itself multiple times on the very first day becoming smarter and smarter (the intelligence explosion). The programmers try to keep their creation "boxed in" so it can't create its own goals. They instruct the software to use its skills just to make some money and nothing else. After Prometheus effortlessly earns billions of dollars, it outsmarts its creators and establishes an apparently benevolent world government. The apparent benevolence is followed by a question mark. Tegmark asks what next? The story is about as realistic as a James Bond movie, but the part about being outsmarted by a computer still rings true.

The underlying idea is called "instrumental convergence"—the tendency for intelligent agents to develop unbounded instrumental goals such as self-preservation and resource acquisition. The conclusion is that apparently harmless goals can act in surprisingly harmful ways.

Convening conferences and writing white papers about the perils and promise of machine learning and artificial intelligence has become a cottage industry. There is a modest-sized community of scientists, philosophers, and intellectuals who are the most urgent in their calls for action. They see runaway artificial intelligence as a looming existential risk to humanity. Nick Bostrom's Future of Humanity Institute at the University of Oxford and Tegmark's Future of Life Institute in Boston are particularly active, and as their names imply, they take these issues very seriously. There are others, among them Jann Tallinn's Centre for the Study of Existential Risk (CSER) at the University of Cambridge, the independent Machine Intelligence Research Institute at the University of California, Berkeley, and Stuart Russell's Center for Human-Compatible Artificial Intelligence also at Berkeley.

Other working groups and white paper teams working on artificial intelligence, however, tend to settle on more neutral ground and suggest that perhaps the killer robot scenarios are distracting and a bit overblown and that the numerous near-term problems of autonomous vehicles, displaced employment, lethal weapons systems, cybersecurity, and privacy regulation need attention first. Typical examples are the Brookings Institution's "How Artificial Intelligence Is Transforming the World" and Stanford's "Artificial Intelligence and Life in 2030." These reports emphasize the distinction between specialized, single-purpose AI applications and the ultimate holy grail of human-equivalent, all-purpose artificial general intelligence (AGI) and focus primarily on the former.

Both the administrations of Barack Obama and Donald Trump convened meetings and drafted white papers on artificial intelligence. Obama's remarks in August 2016 characterize this middle ground:

> The way we think about AI is colored by popular culture. There's a distinction, which is probably familiar to a lot of your readers, between generalized AI and specialized AI. In science fiction, what you hear about is generalized AI, right? Computers start getting smarter than we are and eventually conclude that we're not all

> that useful, and then either they're drugging us to keep us fat and happy or we're in the Matrix. My impression, based on talking to my top science advisers, is that we're still a reasonably long way away from that. It's worth thinking about because it stretches our imaginations and gets us thinking about the issues of choice and free will that actually do have some significant applications for specialized AI, which is about using algorithms and computers to figure out increasingly complex tasks. We've been seeing specialized AI in every aspect of our lives, from medicine and transportation to how electricity is distributed, and it promises to create a vastly more productive and efficient economy. If properly harnessed, it can generate enormous prosperity and opportunity. But it also has some downsides that we're gonna have to figure out in terms of not eliminating jobs. It could increase inequality. It could suppress wages.

August 2016 represented the waning months of the Obama presidency, and he was in particularly good spirits and a pensive mood when he sat down with MIT's Joi Ito and *Wired*'s Scott Dadich. When the supersmart-computer-takeover scenario came up explicitly, Obama laughed and remarked: "And you just have to have somebody close to the power cord. Right when you see it about to happen, you gotta yank that electricity out of the wall, man." Pulling the plug was precisely the solution Alan Turing had originally suggested when these scenarios where first addressed. The difficulty, of course, is that there will be no one plug to pull, no singular machine, but rather an immense network of interacting and learning machines and databases.

So where does the model of evolutionary intelligence come out on this controversial and important question?

Evolutionary Intelligence and Existential Risk

Mary Shelley's moral tale of Dr. Frankenstein's hubris is seductively powerful. Invention involves risk. So, as it has been true of nuclear power and genetic modification, among other scientific domains, risks associated with artificial intelligence and machine learning need to be carefully monitored. The debate is useful. The dramaturgy, even if overdone, does draw in public attention. But I have become persuaded that there will be several thousand intermediate risks that merit attention, serious research, and policy analysis before anything like the Terminator takeover scenario appears on our horizon. I am ruefully sympathetic with Nick Bostrom, who in a particularly cynical frame of mind noted that humans may be more likely to be responsible for our collective demise well before machines would get the chance.

Since I view these developments as largely inevitable, I am skeptical that prohibitive policies could possibly succeed. Private and public investments in AI research and development exceed $50 billion annually and are growing steeply as countries and companies come to define this area as key to competitive advantage. The challenge, then, may best be put to work toward the alignment of machine goals and agreed-on human goals. My personal estimate is that if we encounter unambiguously malicious software code as part of an AI system, 90 times out of a hundred it will have been put there purposely by humans, and 10 times out a hundred it will represent a relatively innocent and unintended mistake. I'm not aware of any malicious code attributable to demonstrable machine self-programming "self-interest." But it's not impossible. That's why working on preventive measures to anticipate and preclude this ultimate potential makes sense.

Why am I not more concerned about HAL taking over?

If we can build self-programming machines that can invent self-interested outcomes, then we can build self-programming machines with the express and explicit purpose of countering them.

The master computer scenario implies some sort of singular computational entity in charge of and controlling all others. The logic of networked intelligence is likely to create a much more complex pattern of distributed intelligence. The notion that these distributed intelligent agents would secretly plot and collaborate to overthrow their human creators is, in my view, an ultimate expression of anthropomorphic projection.

The prototype we are exploring here of evolutionary intelligence is explicitly modeled on alignment of goals. EI is not a force of independent agency; it is an intelligent portal to networked resources under the control of the individual. It is an adviser rather than a decision maker.

Is there still a significant risk scenario we need to confront?

In my view, there is. It is the hijacking of these resources of intelligent computation by individuals and institution bent on malevolent goals. Hijacking is a strong metaphor. It implies pirates on the high seas or a criminal gang taking control of a plane. Hijacking in modern society tends to be more subtle, gradual, and public, with successful public relations efforts to conceal the malevolence. This issue will dominate our final chapter.

The Elusive Goal of Transparency

There is an extremely promising strategic policy to address these various dragons. A real general-purpose dragon killer. The policy is identified, broadly, by the term "transparency." If a search engine is biased, nudging you toward products sold by a corporate partner, then that bias should be testable and demonstrable. Search engine companies are very sensitive to public evidence that they may be purposefully misleading their clients. If it is empirically shown that a financial algorithm is geographically and racially biased, we should insist that those who choose to use it are legally liable. If a company obscures or misrepresents its privacy policies, let its potential customers know. If a machine learning algorithm is writing code for itself that may include malicious agency, carefully review the code. If the code is complex and difficult to evaluate, put another intelligent algorithm to work examining the source code and ferreting out potentially problematic agency.

The enemy of transparency is propriety—that is, the protection of secret algorithms on the grounds that they represent a trade secret or a protected proprietary process. A classic case is COMPAS software, a criminal justice "assessment system" widely used in state courts around the United States and famously in Florida. The acronym stands for Correctional Offender Management Profiling for Alternative Sanctions. The system attempts to predict the likeliness of a criminal reoffending and is used as a guide in criminal sentencing. The problem, according to a ProPublica study, was that African American defendants were almost twice as likely to be misclassified with a higher risk of reoffending compared to their white counterparts—45 percent for blacks, 23 percent for whites. Northpointe, which owns COMPAS, won't make its formulas public because the system is proprietary. Fortunately, however, the public use of the system creates public data on sentencing that is subject to analysis and critique. Northpointe disputes the ProPublica study as flawed and continues to sell its system profitably.

Typically, the problem in machine learning systems is either bias in the training data or missing data used to build and tune the system. Amazon tried to streamline its hiring process with algorithms that apparently reproduced a pro-male bias in hiring that was manifest in its previous human-based system. Similarly, the National Institute of Standards and Technology reported facial recognition systems used by police and federal agencies made errors in identifying African American and Asian faces more frequently than Caucasian faces because of inadequate training data. These are important and persistent problems, but they are correctable with an appropriate investment of time and energy.

Because the development and application of these technologies happens so dynamically, it is not practical to have some sort of federal or nonprofit entity testing machine learning algorithms for bias and error. The existence of such an entity or perhaps a breadth of entities would be welcome. But the best chance for effective monitoring is competition among commercial entities that carefully and continuously monitor what their competitors are up to. If Apple or Microsoft or Google or Amazon or a smaller start-up finds evidence of a competitor covering up errors and biases, they have every incentive to make it public.

We started this chapter by introducing the dragon metaphor. When exploring the unknown, we are prone to imagine dangers lurking around the corner. In the case of evolutionary intelligence, we have the advantage of a depth of experience with its technical forebearers. Algorithm bias, overdependence on convenient technologies, privacy challenges, and technical accidents with potentially serious consequences provide us sobering lessons for the road ahead. But perhaps some general conclusions can be drawn reviewing EI as a particular case of technical convergence.

First, a strategy of prohibition is not likely to be successful. Given the widely accepted belief that corporate and nation-state investment in various dimensions of artificial intelligence is going to be profitably rewarding, such a policy is not likely to win support or to be practically enforceable. Unlike nuclear weapons, where the radioactive physical characteristics of the technology can be monitored in various ways, there is no reasonable prospect of policing code-writers and systems designers.

Second, a strategy of encouraging best practices and safeguards *is* likely to be successful. The ultimate metrics of the success of variants of evolutionary intelligence are that those who use it judge it beneficial. Google search is one of the most highly rated commercial services in recent history. Google and its search subsidiaries such as YouTube dominate about 90 percent of search activity worldwide. There are plenty of competitors, but people continue to be satisfied and conclude that Google helps them find what they are looking for. Rumors of search bias and privacy issues pop up from time to time, but Google has every incentive to take those challenges seriously. Market competition is the best medicine, the most promising by far.

Third, a strategy of transparency is important. In this case, commercial and nation-state players may prefer to obfuscate self-serving mechanisms in the algorithms they promote when and if they can get away with it. Here again, competitive players will be the best and most attentive system testers and alarm ringers. That goes for nation-states as well as commercial entities. If the code is proprietary, test the behavior of the black-box system. If hacking is involved, look for the telltale signatures of the hacking techniques. There are going to be active malicious players online. EI systems will serve as modern-day security and antivirus systems do in laptops and smartphones. EI will have the capacity to monitor outcomes and protect against the bad guys.

Fourth, and finally, the computers-take-command scenario is an important question, but not an urgent one. There is much that can be done under the strategies of best practices and transparency in the meantime. But I'm with Stanford's Andrew Ng—worrying about the rise of evil killer robots is like worrying about overpopulation and pollution on Mars well before we've started out on a visit.

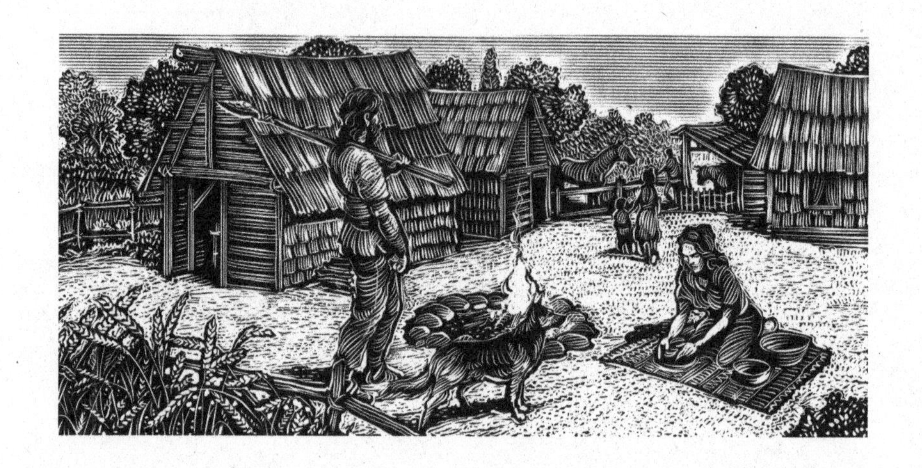

5
Evolutionary Intelligence in Historical Perspective

Many modern observers use the term "revolutionary" quite casually. The term helps to draw attention to your topic. Perhaps we have become overly accustomed to the term after hearing about multiple revolutions in dishwashing detergents. But I propose to take the concept very seriously. Evolutionary intelligence (EI) has revolutionary potential.

I'm going to make the case that there are four fundamental inventions in the history of humans on earth that help put the revolutionary character of EI in perspective. The four are *language, land, leverage,* and *literacy*. Think of each as a human-invented technology.

By language, I refer to the capacity to convey concepts and relationships beyond the simple use of grunts and gestures. The date is difficult to pin down, but the experts believe that what we would consider modern speech is only about 100,000 years old. That's a small fraction of humanoid existence on earth. It is a bit disconcerting to realize that we have probably been just grunting, making animal sounds and pointing for most of our time on this planet.

By land, I mean the historical process of settling down in one place rather than wandering around in search of berries, tubers, and small game. We have been settled on land for only 10,000 years. It was made possible by the discovery of agriculture and the domestication of animals. It took us quite a while to figure that out.

By leverage, I identify the transition from using tools with human or animal power to mechanical power—the use of steam, internal combustion, and electricity. The concept of leverage corresponds to what is typically called the Industrial Revolution.

By literacy, I don't mean written language or printing presses, although both are important. I mean mass literacy—the capacity of the average individual to access the culture around them directly through reading and writing. Mass literacy begins with the mass education that accompanied the leverage of the Industrial Revolution in the nineteenth century and has now spread to most of the globe. The printing press of the fifteenth century created books that only a tiny fraction of the population could actually read.

These big four—language, land, leverage, and literacy—set the stage for the widely acknowledged current and ongoing transition of the fairly well understood digital revolution as we shift from an industrial to an industrial/information economy. Those four revolutions, plus the current expansion of computational intelligence, I will argue, lead powerfully and inexorably to the next fundamental change that will, in turn, equivalently change the nature of our collective existence.

First, Language—The Enhanced Capacity to Share Ideas

The first big thing that fully defined our humanness and distinguished us from our primate forbearers was the development of language—our capacity to speak with each other and coordinate our activities beyond the level of facial expressions, grunts, and gestures.

Perhaps a thought experiment would be useful.

Ponder the question, Can we think without words? Put the book down and reflect on that for a minute.

What have you concluded? People diverge on this one. We typically engage language in our dreams. It is likely that the exercise of imagining about thinking without words probably stimulated a few words and phrases in your head. It would seem that some basic thoughts or memories are possible without a linguistic label. Something tastes good or smells bad. One can imagine the feeling of fear or the feeling of happiness being clearly understood even if either lacked a linguistic label. But what about more abstract concepts about logical relationships, quantities, or unfamiliar events?

Probably the best example of the language-thought interaction is perception of quantity. Physicist George Gamow published a popular science book in the 1940s with the title *One, Two, Three . . . Infinity*, which emphasized this point. Some cultures had no words or concepts for any quantity larger than three. For those cultures, it was simply "many." Such a linguistic/cognitive limitation would appear to be quite a constraint on the economics of trade and exchange.

Since there are virtually no cultures currently in existence without some form of speech, we have no natural control group of the linguistically deprived to explore this question very far. There have been studies of deaf children, pre-language infants, and adults with global aphasia who have lost the capacity for speech. In each case, the evidence is mixed. Some form of thinking can take place without language, but language is intimately intertwined with how we perceive and respond to our environment. The fact that there are virtually no cultures on earth yet discovered without some form of speech is worthy of note. Culture, speech, and language are bound up together.

Could you convey a moderately complex narrative to someone without the power of language? Think charades. Not very easy. That is what makes charades fun.

Alternatively, imagine life within a family or small tribal group without the capacity for language. Put the book down and reflect on that for a minute.

Well, what do you think? A lifetime playing charades? My effort here is to draw attention to how critically important language has become for making us human. It is a truly transformative creation. And the best research indicates that linguistic systems among humans have only existed for between 160,000 and 80,000 years. That is stunning. For the first 200,000 to 300,000 years of physically modern human beings on earth, it was life in caves, hunting and gathering, small tribal groups, and variations of charades with gestures and onomatopoeics.

We turn at this point to psychologist Michael Tomasello's provocative intellectual puzzle. If we trace the evolution of *Homo sapiens* in Africa, the process follows a steady Darwinian process of gradual differentiation of primate species and selective survival over millions of years. Then something very dramatic happens. Tomasello puts it this way:

> The basic puzzle is this. The 6 million years that separates human beings from other great apes is a very short time evolutionarily, with modern humans and chimpanzees sharing something on the order of 99 percent of their genetic material—the same degree of relatedness as that of other sister genera such as lions and tigers, horses and zebras, and rats and mice. Our problem is thus one of time. The fact is, there simply has not been enough time for normal processes of biological evolution involving genetic variation and natural selection to have created, one by one, each of the cognitive skills necessary for modern humans to invent and maintain complex tool-use industries and technologies, complex forms of symbolic communication and representation, and complex social organizations and institutions. And the puzzle is only magnified if we take seriously current research in paleoanthropology suggesting that (a) for all but the last 2 million years the human lineage showed no signs of anything other than typical great ape cognitive skills, and (b) the first dramatic signs of species-unique cognitive

> skills emerged only in the last one-quarter of a million years with modern *Homo sapiens*. . . .
>
> There is only one possible solution to this puzzle. That is, there is only one known biological mechanism that could bring about these kinds of changes in behavior and cognition in so short a time—whether that time be thought of as 6 million, 2 million, or one-quarter of a million years. This biological mechanism is social or cultural transmission, which works on time scales many orders of magnitude faster than those of organic evolution.

Given that, as a species, we humans are not very strong or very fast (or well equipped for cold climates), our capacity to hunt collectively and with tools and clothing and to share and refine strategies for gathering wild seeds, grasses, nuts, seasonal vegetables, roots, and berries predicated on language has clearly been key to our survival.

Language as the basic form of social and cultural transmission is the first major revolution because it gave us the capacity to understand the world beyond our immediate experience, to build on the experience of others, and to pass that on to our progeny.

Language may take the prize for the biggest big thing. What a change in the character of human existence, yes? It turns out, big thing number two comes very close.

Second, Land—Settling Down after Millennia of Nomadic Hunting and Gathering

As far as we know, life as a hunter-gatherer was largely a desperate nomadic search for food. Life on the run meant little time for much human contact outside the immediate family. Too many people too close together meant the immediate exhaustion of the food supply. This fact limited population growth at several levels. First, of course, came starvation and near-starvation conditions. Second, experts speculate that children were spaced out because a mother could carry only a single child and toddlers were a major impediment to mobility.

The good news for the genetic strains of humanity that sustain us now is that these tough times meant that the survivors are a sturdy stock. A very difficult existence for hundreds of thousands of years, perhaps 10,000 successive generations—it is difficult to imagine.

At this point, about 50,000 years ago, physically modern humankind had now moved beyond Africa and spread around the world hunting and foraging. And then, 10,000 years ago, something happens. It happens pretty much everywhere. It happens in multiple locations independently. Scholars wonder whether the word was spread around by travelers and traders. Humans discovered the next big thing—settling down in one place and farming and domesticating animals. Families grew into tribes, villages, and ultimately cities. The increased availability of food meant there was now time to develop tools and build shelters and religious monuments and for an elect few to emerge in positions of authority as leaders, priests, and warriors. Historians call this the Neolithic revolution. It is a competitive candidate for the biggest big thing ever. Rather than moving out with whatever we could carry, we settled down, built houses and villages, and, by necessity, developed the social norms that make larger-scale collective human life possible. Land had new meaning. It was no longer what you traversed on your way somewhere else. It was where you lived and what you defended.

This isn't a history book. It is a book about a dramatic development in our future. But at this point it is important to draw a few insights from these previous social-physical revolutions to better understand the next. Scholars, as is their wont, get into heated debates about the when, where, who, and whys in this period. The debates are heated because we have such fragmentary evidence and, of course, no written record whatsoever. Here are the lessons I would like to draw from these debates. As best we can tell, as the Neolithic revolution was going on, human beings were only vaguely aware of its significance for their lives. It did, after all, take about 2,000 years to come to full development. Some scholars argue, in fact, that the effort of these early farmers was not to set up a revolutionary new way of life but to protect the lifestyle that they had known that was challenged by changing environmental conditions.

Jared Diamond, for example, offers an observation about the Neolithic revolution:

> What actually happened was not a discovery of food production, nor an invention . . . food production evolved as a by-product of decisions made without awareness of their consequences. They half-knew they were on the road to the eminently desirable goal of becoming farmers. . . . In fact, as the regional case studies have shown, it seems more likely that in many instances foragers were attempting to preserve their way of life at a time of stress, rather than deliberately seeking to transform it.

So we can understand the Neolithic revolution to be the product of hundreds of thousands of individual micro decisions about seeds, irrigation, and figuring out how to manage reluctant animals for wool, eggs, milk, and slaughter. One intriguing scenario (and, of course, we can only speculate) is that hunters chased a group of goats into some sort of canyon and figured out how to keep them there. The abundance of available meat meant that the herd could be allowed to multiply and continue to provide a bountiful supply.

The transition was difficult and complex. Ironically, the early lives of settled farmers and herders were even more demanding and difficult than the life of the hunter-gather as crops failed, animals ran off, and denser communities led to bouts of epidemic disease, drought, flooding, and fire. So, despite the increased efficiency of food production, the population would not grow dramatically until some years later. The transition from forager to farmer happened independently about 8,000 to 12,000 years ago as the last ice age receded and sedentary life was possible in nine documented areas of the world—the Near East, North China, South China, sub-Saharan Africa, South Central Andes, Central Mexico, Eastern North America, Highland New Guinea, and Amazonia. It is an impressive list.

In addition, scholars posit that farming developed as well by means of cultural diffusion in Northwestern Europe, Southwestern North America, and Japan about the same time.

Researchers characterize the transition as a seemingly irresistible process. Peter Richerson and colleagues describe the stark social dynamics as follows:

> Once a more productive subsistence system is possible, it will, over the long run, replace the less-productive subsistence system that preceded it. The reason is simple: all else being equal, any group that can use a tract of land more efficiently will be able to evict residents that use it less efficiently. More productive uses support higher population densities, or more wealth per capita, or both. An agricultural frontier will tend to expand at the expense of hunter-gatherers as rising population densities on the farming side of the frontier motivate pioneers to invest in acquiring land from less-efficient users.

The archaeological evidence from this era supports this argument. The abundance of food would lead to complex city-state civilizations in Central and South America, Babylonia, Sumer, the Assyrian Highlands, the Indus Valley, China, and ultimately Persia, Greece, and Egypt. The resources and available time that led to creating art, written language, and religious pageantry also led, as we know well, to armies, military technology, and war. Tribal and ethnic conflict, it would appear, was as inevitable as the agricultural technologies that permitted it. It is not always widely cited but remains a significant historical lesson—a challenging aspect of hardwired (and thus unchanging) human nature to which we will be returning in further chapters.

Third, Leverage—The Dawn of the Industrial Revolution

For 99.9 percent of human existence on earth, we got it done with animal power. We relied on our own muscles and the strength of horses, oxen, and mules. We were inventive, so we had a few specialized uses of wind and water power for transportation and milling. Now we have the leverage of machine power. Do you drive a horseless carriage with abundant horsepower? Of course you do. What else could we call these things? Imagine that. A carriage without a horse. The term "automobile" has the same basic connotation—an awesome carriage that propels itself.

Given how central and taken-for-granted car culture and industrial infrastructure are to modern societies around the world, it is hard to imagine that these mechanical devices have only been with us for a century. Karl Benz built the first of what we would recognize as a reasonably modern internal combustion–based motor car in 1885 in Mannheim, Germany. The beginning of the automotive revolution started fitfully. Benz sold only about five cars a year for the next five years. It wouldn't be until 1908 with Henry Ford's assembly line–produced Model T (with a new car coming off the line every 15 minutes) that automotive technology became accessible to the masses.

As the history books and encyclopedias report breathlessly:

> The industrial revolution marks a major turning point in history; almost every aspect of daily life was influenced in some way. In particular, average income and population began to exhibit unprecedented and sustained growth. Some economists say that the major effect of the industrial revolution was that the standard of living for the general population in the Western world began to increase consistently for the first time in history. This transition included going from hand production methods to machines, new chemical manufacturing and iron production processes, the increasing use of steam power and water power, the development of machine tools, and the rise of the mechanized factory system. The growth of modern industry since the late eighteenth century led to massive urbanization and the rise of new great cities, first in Europe and then in other regions, as new opportunities brought huge numbers of migrants from rural communities into urban areas. In 1800, only 3 percent of the world's population lived in cities, compared to nearly 50 percent today.

In retrospect, we have come to see it as a timely convergence of perfecting iron and steel manufacturing and the invention of reliable and powerful steam engines. Railroads and steamboats revolutionized transportation. Steam power made the large-scale textile and consumer-goods factories practical and economically efficient.

Industrialization, as we know well, had many unanticipated and highly negative social effects. Life in the newly teaming cities was characterized by long hours, meager wages, dangerous working conditions, exploitation of child labor, crowded tenements, and many sanitary challenges. Epidemics of cholera, typhoid, and typhus were frequent and devastating. Fire was a constant threat. Experts studying the transition from hunting and gathering to agriculture and animal husbandry made two points: (1) Those generations that were part of the revolutionary transition did not see it as such, and (2) the transition was wrenching and difficult for many generations before new methods of food production and models of social organization became refined and routinized. Both points were certainly true of the Industrial Revolution. We noted in chapter 1 that almost a full century of industrialization had taken place before the summative term "the Industrial Revolution" came to be used. And, it could be argued, we have not even yet fully come to grips with its iniquitous effects on social life and the environment.

Sociology as a discipline is in large part a child of the Industrial Revolution. The founding scholars recognized the dramatic social changes and, each in their own way, felt that the disciplinary traditions of history, political science, and economics were inadequate to the important task of making sense of this revolution in technology and human organization. Ferdinand Tönnies and Max Weber, for example, developed the analytic dichotomy of *gemeinschaft* versus *gesellschaft* as fundamental principles of social organization. Gemeinschaft characterizes the basic organizing basis of the small preindustrial community where almost everybody knew everybody else and relied on interpersonal trust. Gesellschaft, in contrast, characterizes the large industrial city where, for the most part, strangers conduct business by relying on bureaucratic rules and large institutions. Marx emphasized the alienation and anomie of the assembly-line worker in contrast to the local artisans and cobblers who were skilled and proud of their craft, knew their customers, and prized their neighbors' and customers' trust and respect. Technology and social structure are intimately intertwined.

Fourth, Mass Literacy—The Educational Revolution and Dramatic Expansion of the Middle Class

I have a clear recollection of traveling near the Rhine River in Strasbourg, France, where the river serves as the French–German border. We came upon a dramatic statue of Johannes Gutenberg, and I insisted that my family take a picture of me posing in front of it. If you spend your life studying the technologies of communication, you just have to get that picture. Gutenberg is "the man." Movable type. The printing press. The veritable invention of mass communication.

Wait a minute. Was that Strasbourg or Vienna? Or Mainz, his hometown? Was it Lodz? Hannover? Frankfurt? Perhaps my recollection is not so clear. Wikipedia, as it turns out, lists 43 such statues and memorials around the world.

My point is that Gutenberg and his press are easy to pinpoint. A single person. A concrete device. A specific year—1455—when he printed his first bible. Brilliant inventor-hero. World-changing invention.

As is often the case, the full story is much more complicated. The world's first movable type printing technology for printing paper books was made of porcelain materials, and it was invented around AD 1040 in China during the Northern Song dynasty. Bi Sheng (990–1051) turns out to be "the man." The first actual book printed with movable metal type was printed in Korea in 1377. The authorities at the time in both China and Korea were not so keen on the potentially unsettling effects of this worrisome technology, so its use was highly constrained and the subject matters to be printed were carefully monitored. Gutenberg gets the lion's share of the credit because he was in Europe, and his tinkering with metallurgy and modified wine presses happened right around the beginning of the Enlightenment. Printing press equals social and cultural revolution? Cause and effect? We can celebrate the invention, but it is hard to argue that printing presses had revolutionary power when only one or two percent of the population was literate. The real revolution, in my judgment, came more slowly and represented a growth in universal education, mass literacy, literacy-based occupations, and the growth of an expansive middle class.

When only a tiny percent of the population can read, it turns out that the economics of opening a print shop are, well, tenuous. Around 1439, Gutenberg lost a lot of money in manufacturing polished metal mirrors that were supposed to capture holy light from religious relics. So he tried using his skill in metallurgy to try printing. By 1450, he had a workable press and printed poetry. In 1455, Gutenberg completed his famous 42-line folio bible and printed 180 copies. Copies sold for 30 florins each, which was equivalent to about three years' wages for an average clerk. Gutenberg had borrowed heavily to support this enterprise, and his creditors sued him for what they characterized as misappropriation of funds and took over his print shop and his inventory of bibles. Gutenberg's successors also struggled financially. It would take some time for a print culture to evolve. As an impoverished old man, Gutenberg did receive some recognition for his technical accomplishments in the last decade of his life as a "gentleman of the court," and he received an annual allocation of grain and wine. It was not to be the riches of a Bill Gates or Steve Jobs. No one had thought to arrange for a portrait of him while he was alive, so virtually all of the 43 Gutenberg statues and plaques are imaginings of what he probably looked like. It would be almost four centuries before the full cultural implications of his brainchild would be manifested in the steam-driven industrial roll presses and daily newspapers of the 1830s.

The nineteenth century in the West and the twentieth century throughout the rest of the world witnessed the spread of universal or near-universal literacy. In 1800, the world literacy level was below 10 percent. The graceful curve upward reaches about 85 percent in the present day for the world as a whole and 95 percent for the industrialized world.

Reading and writing for the mass population, according to historian Carl Kaestle, brought about significant changes:

> [It] allows the replication, transportation, and preservation of messages, and it allows back-and-forth scanning, the study of sequence, deliberation about word choice, and the construction of lists, tables, recipes, and indexes. It fosters an objectified sense of time, and it separates the message from the author, thus "decontextualizing" language. It allows new forms of verbal analysis, like the syllogism, and numerical analysis, like the multiplication table. The long-range developments made possible by this technology have been profound, leading eventually to the replacement of myth by history and the replacement of magic by skepticism and science. Writing has allowed bureaucracy, accounting, and legal systems with universal rules. It has replaced face-to-face governance with depersonalized administration. On the other hand, it has allowed authorship to be recorded and recognized, thus contributing to the development of individualism in the world of ideas.

The printing press signals the technical affordance of mass communication. Initially, it empowered the intellectuals and Encyclopedists. Jean-Jacques Rousseau read about what David Hume and Adam Smith were thinking. Immanuel Kant and Arthur Schopenhauer in turn read Rousseau and took his thinking further. But it would take a while for these enlightenment ideas to start to converge with popular culture and the daily lives of the citizenry. The universal capacity for reading and writing is the ultimate social and cultural result of that affordance.

Today more than half of the world's population is 'middle class.' Some argue it is perhaps the most dramatic demographic shift ever witnessed. It's easy to forget that the middle class barely existed for most of modern history. Brookings scholar Homi Kharas notes: "There was almost no middle class before the Industrial Revolution began in the 1830s. It was just royalty and peasants." Being middle class is a big deal. The growth of the middle class beyond the industrialized West is still in dramatic transition. As recently as 1975, 60 percent of the world still lived in extreme poverty.

The significance of the revolution of the middle class is captured in Max Weber's notion of improving "life chances." Weber argues that opportunities in this sense refer to the extent to which one has access to not just food, clothing, and shelter but less tangible resources such as education, health care, and potentially fulfilling employment. Life chances correspond to the individual's ability to satisfy primary needs, have a career and obtain, as psychologist Abraham Maslow would put it, self-actualization. The emergence of the middle class as the culmination of industrialization and mass literacy embodies the dramatic transformation of daily life—for the better—for millions of people.

What Is Revolutionary?

Our abbreviated review of the developmental history of language, land, leverage, and literacy characterizes each as revolutionary. Global. Life-changing. Irreversible. They fundamentally change the way we subsist and how we spend our time. They transform how we organize ourselves socially. Each identifies a transformative technology. The transformations involve a mix of changing how we interact with our physical environment and how we interact with each other.

These are not case studies of technological determinism. The technologies may provide affordances the originating culture has no use for. The printing presses of ancient Korea and China went unused. They were viewed as potentially threatening to the existing social order and royal succession. Scribes and priests in the ancient Middle East made it a point not to encourage mass literacy for fear of losing the power and status that their unique skills afford and ultimately the prospect of losing their very jobs.

Let's be clear. This is setting a very high bar for the case about evolutionary intelligence in historical perspective. Something revolutionary, not just an incremental improvement in how things are done. Not just a shift from hieroglyphics to phonetic alphabets. Not just a shift from black and white to color. Something fundamental.

And Finally—The Digital Revolution

So far in these pages we have made note of the fact that the Industrial Revolution was not holistically recognized as such until about 100 years of industrialization had taken place. This is clearly *not* true of the digital revolution. It is certainly well argued over, if not necessarily well understood. Observers have been agonizing about the awkward mix of positives and negatives (mostly positives?) of the ubiquitous computer screen on our desk or equally predictable smartphone in our hand. The digital revolution has been spawned by a special class of industrial inventions we call computers. Unlike the steam engine, the internal combustion engine, and the nuclear engine, the computer processes ideas rather than things. Unlike the rail, highway, and air transportation infrastructure, the Internet moves bits rather than atoms. What is it about this particular development among the other hundreds of thousands of inventions of the Industrial Revolution that leads us to identify its social, cultural, and economic impact as truly revolutionary? The answer is that this technology "knows" something about the subject matter it processes. Andy Lippman, my colleague from the MIT Media Lab, calls this phenomenon "intelligence in the channel." In the pages ahead, we will explore why it continues to be so life-changingly significant.

Think about most inventions of the Industrial Revolution as, well, akin to a wood chipper. A wood chipper knows one thing. Whatever you put in the hopper is going to get ground up. Although it sounds odd to say, the wood chipper has no concern about what you put in the hopper. These are the products of the industrial age. The steamship happily heads straight for an iceberg unless otherwise instructed. A finely engineered, high-powered rifle will as soon shoot a person as a target.

Computers are different. Ask a computer to divide by zero. The computer will object. Wait a minute. No can do. Type "adn," out comes "and." Type a zip code and the computer fills in the city and state. Take a picture with your smartphone. If the scene is bright, the exposure setting has been adjusted appropriately. Digital storage takes note of where the picture was taken based on data from the phone's GPS receiver. The phone may run a facial recognition routine and duly make note of the names of those that are familiar faces.

This is intelligence in the channel. Digital technologies know something about what they are doing. They don't know everything. They aren't self-aware (yet). But they are designed to respond differently depending on the input they receive. And more than just responding differentially based on predesigned codes, they can "learn" from experience.

Many of us have heard about Herr Gutenberg's printing press, Alexander Graham Bell's telephone, and Philo Farnsworth's television set. Fewer of us, however, have heard about the Turing machine, a machine never built. Alan Turing's fundamental idea was stunningly simple and stunningly important.

According to legend, in the summer of 1935, Turing was lying in a field in Grantchester, a rural village just south of Cambridge University, when an idea came to him. The idea captured his imagination and his work life and resulted in a publication the following fall. The work had the intimidating title "On Computable Numbers, with an Application to the Entscheidungsproblem" and appeared in the *Proceedings of the London Mathematical Society*. Entscheidungsproblem is German for "decision problem," and it represented one of a famous set of mathematical puzzles proposed in the 1920s about the solvability of specified equation sets. To answer this question, Turing invented a machine—a device we have now come to call the Turing machine. The machine itself could never be built because (among other things) its central recording tape was described as "infinitely long." But like Einstein's thought experiment of riding along on a light wave to visualize the idea of relativity, Turing's machine was a theoretical machine, designed not to be built but to make a point.

The machine operates on an infinite memory tape divided into discrete cells. The machine positions a reader over a cell and scans the symbol there. The symbol is then compared with a lookup table of user-specified instructions. The instructions (what we would call computer code today) may dictate that the machine move on in a new direction or perhaps overwrite the symbol. That's pretty much it.

That sounds simple, even simplistic. But there is a powerful concept hidden therein. The concept is conditionality. What the computer does next depends on what symbol the computer reads and its corresponding instruction set. This is the fundamental computer coding logic of an "if statement" and a "do loop." The machine moves to the next step in the instruction set (the looping instructions) only if the read symbol satisfies the conditions specified in the instruction set. Does the password entered by the user correspond to the password on file? If not, display message "Username or password incorrect. Passwords are case sensitive. Perhaps the caps lock is on. Try again."

There is no shortage of theorists, pundits, critics, journalists, and a few scientists writing books about what the digital age means for all of us: the computer age, the information economy, the Internet-connected world, the new media—many names for basically the same historical phenomenon—the shift from an industrial/manufacturing economy to one based on information technologies. Jobs traditionally associated with the middle class such as assembly-line work are declining as automation improves productivity. Jobs for mind workers such as engineers, attorneys, teachers, scientists, designers, programmers, and executives are increasing. The digital age starts in the 1940s with room-sized computers built with vacuum tubes. As transistors replaced tubes and as microchips got smaller and more powerful following Moore's famous law (doubling their capacity for computation every 18 to 24 months), big computers (1950s) were replaced by minicomputers (1970s), which in turn were replaced by personal computers (1990s) and smartphones (2000s). Even the term computer itself seems archaic as computational intelligence becomes ubiquitous in light switches, thermostats, traffic lights, cars, and airplanes.

Does the idea of self-driving cars and computer-driven airplanes strike you as a little scary?

Me too.

One ponders the stories of pilots doing battle with an errant flight-control computer in freshly manufactured Boeing 737 MAX aircraft and losing those battles, resulting in two crashes and the loss of 346 lives. This theme of who is ultimately in control here—the computers or the humans that designed the computers—is one we will return to. It is the Frankenstein narrative of technologies we do not fully understand and cannot fully control. It resonates deeply with all of us from the industrial through the digital age. The movie snippet described at the beginning of this book from *The Terminator* motion picture franchise sustains this theme in the dystopian image of the ultimate battle-to-the-death between humans and machines. Because these movies have been very profitable, the narrative gets rewound and retold with variation many times. We clearly have an appetite for this moral tale. We usually seek a happy ending in our storytelling and, of course, in the real world—we seek to design machines that are on our side, programmed to do no harm, programmed to help. Is this not possible?

The best lessons for understanding these delicate dynamics of control, of technology in social and economic context, in my judgment, can be found in the precursors: the half-developed, half-thought-out early prototypes of EI that can be found all around us now.

6
Evolutionary Intelligence Is Already Here

Sherlock Holmes's sidekick Dr. Watson was a most convenient narrative foil. "Elementary, my dear Watson," Holmes would explain with partially suppressed pride as the good doctor would ask Holmes the simple questions that had likely puzzled many readers as well. The clues were plainly obvious to Holmes's acute perception but usually unseen or misunderstood by the rest of us. But detective Holmes pondered only fictionalized puzzles. In this chapter, we explore stories of some real-world prototypes of evolutionary intelligence (EI) much closer to our theme of historical revolution—true stories, no less. Well, mostly true. Important historical events are often argued over and remembered with different details by the participants. My argument is that the revolution is upon us already, although perhaps, like the Industrial Revolution, we just don't see it as such. Why not?

Our first story starts on a December day in Silicon Valley in 1979. The scene is set in a conference room at Xerox Palo Alto Research Center (PARC), the famous research and development (R&D) facility of the then dominant office products company. PARC was just up the hill from Stanford University and shared the famous research culture with the university and the many start-ups on the peninsula. The "suits" back at Xerox headquarters in New York had an inkling that computer screens might replace paper in the office and concluded (correctly) that that would spell trouble for Xerox if they didn't get out in front of this threatening trend. And get out in front they did. PARC was well-funded, well-staffed, and informally structured around projects in a way that encouraged dramatic creativity. The team at PARC had built a working prototype of what would become the personal computer with a mouse and a windows-oriented graphical user interface. In the 1970s, communication with computers required the painstaking typing of multiple command lines (no typos permitted).

PARC's test model was called the Alto. It was the product of a decade of hard work. At this point it was still a carefully guarded proprietary secret. The staff at PARC recognized the potential significance of what they had built. For several years, they had traveled back East to try to convince senior management that this could be the future direction of Xerox.

In one video reenactment created many years later, the PARC engineers are sitting at the far end of an immense boardroom table as they earnestly present the Alto and the innovative mouse controller. They were surrounded by gray-haired businessmen in expensive suits frowning with great skepticism. When the presentation is concluded, the mouse is handed to the first board member, who silently passes the mouse along to the next board member, who passes it to the next, and finally to the apparent chairman of the board at the opposite end of the gigantic table. The chairman pauses, frowns, and shaking the tiny prototype says in a booming voice: "You mean to say that the Xerox Corporation is supposed to sell a product called a mouse?" We are not sure if that was what was actually said. The video was a documentarian's visualization of the apparent cluelessness of management made many years later with the benefit of historical hindsight.

We do know that what actually happened was that Xerox headquarters sent two business acquisition guys to talk to the folks at this new, hot start-up down the road from PARC called Apple Computer. Xerox was having trouble competing with Asian competitors who could make similar office products much more cheaply than Xerox. Apple had a reputation for high-quality, low-cost manufacturing know-how. The acquisition guys struck a deal with Apple that they would open the kimono and show them the Alto prototype in exchange for some Apple stock and some (rather vaguely defined) manufacturing process help. What the business acquisition guys didn't do was talk to PARC about any of this, and the folks at PARC hit the roof. For them, it was giving away a decade's worth of work for no obvious benefit. Adele Goldberg, one of the project leaders at PARC, famously declared that she simply wouldn't do it unless directly ordered to do so. And sure enough, they gave her a direct order.

So, in mid-December 1979, a rather tense meeting took place including Larry Tesler and Adele Goldberg and others from PARC and a small team including Steve Jobs from Apple. It is a famous scene described many times since by the original participants at various industry panels and reenacted in several movies and documentaries. Tesler runs the Alto through its paces. The engineers from Apple are excited and full of questions. They had heard about some of these ideas but never seen them actually work. A mouse. A graphical interface with drop-down menus. PARC had a working model of the human–computer interface that we have come to simply take for granted since the Mac and Windows-based PCs of the mid-1980s. Jobs was pacing nervously around the room looking at the Alto and looking away. Suddenly (according to Tesler) he blurted out: "What's going on here? You guys are sitting on a goldmine. Why aren't you doing something with this technology? You could change the world." Jobs himself would later muse: "Within ten minutes it was obvious to me that all computers would work like this someday. They were copier-heads. They had no clue. They grabbed defeat from the greatest victory in the computer industry."

Jobs was right. Xerox tried a few times but never made a dent in the computer marketplace. The folks at PARC understood what they had, and several of them famously ended up working at Apple. The clues were all there. No one had quite put them together. The mouse, for example, had been introduced by Douglas Engelbart from the Stanford Research Institute at the famous mother-of-all-demos at a trade show in San Francisco back in 1968. But nobody had yet put the mouse and graphical interface it controlled together.

So that is the takeaway from our brief visit to a tense meeting room in Palo Alto. Our task now is to put the clues about the future of technology together into a coherent pattern. The task in this chapter is to look closely at a few clues that surround us today. A pattern will emerge. At the end of the chapter you get to say "What is going on here? We are sitting on a goldmine. Why aren't we doing something with this technology? We could change the world."

We reviewed other major social-technical revolutions in the last chapter and saw the same hidden-clue phenomenon—the seeds of the next development abound in the previous era and can be found if you look for them. Protolanguages of various forms were probably ubiquitous in the millennia before full-scale structured languages evolved. Hunter-gatherers would cultivate some limited wild cereals when circumstances permitted. Entrepreneurial industrialists would pursue wind and water power for milling and manufacturing before steam, internal combustion, and electricity introduced a full-scale industrial restructuring. Mechanical analog computers preceded modern digital computation. So where are these precursors of evolutionary intelligence?

Such clues abound. But we will focus on five of them. They represent individual trees, not yet a forest. The clues are miscellaneous, helpful conveniences of electronic intelligence. They all involve a process of communication of some sort—sometimes communication between machines, but mostly among humans.

The Magic of Self-Configuration—Consider the Lowly Chirping Fax Machine

Some younger readers will have only a vague familiarity with facsimile communication (fax for short). It is simply two machines connected by phone lines that accept pages of input at one end and reproduce copies, one page at a time, at the other. Fax machines were nearly ubiquitous in businesses in the 1970s through the 1990s for text and image-based commercial communication. Since phone lines are optimized for audible voice transmission, the machines converted visual scans into data as audible tones. Patterns of low tones and high tones represented zeros and ones. Although many fax machines today remain plugged in and are occasionally used, online digital communication has largely overtaken them. What about the noisy part? Some readers will recall the familiar and rather cheery beeps and chirping tones as one fax machine greets another. That's the tantalizing clue, the audible "Hello, I'm a fax machine, are you?" In this case, it is an early exemplar of machine-to-machine communication.

This is typically referred to as a digital handshake. A bit more is underway than simply mutual recognition of two fax machines on a telephone line. In fact, a lot more. In order to permit efficient transfer, both machines must establish the fastest practical data rate and details like page length and width, data compression format, and scanning units per line. The sending machine might politely ask if the receiver was equipped to reproduce color images. The clue here lies not in the details particular to facsimiles but the underlying logic of intelligent, instantaneous, self-adjusting machine-to-machine communication. The chirps signify a deeper process known as *self-configuration*.

The beeps take place when a phone call is answered. Typically a fax machine doubles as a traditional answering machine. If you start talking, the machine records the audio. If you start beeping, it beeps back to set up the standards for visual-graphic fax communication. So the handshake sets up the modality of communication, audio or visual. The fax machine is 1960s technology, so the audio quality and graphic resolution are pretty primitive. And slow. In 1974, it typically took three minutes to fax a single page across the telephone network. Early fax machines cost thousands of dollars and required special and difficult-to-manage thermal paper. Ultimately, of course, fax communication would be abandoned and replaced by online graphic communication of document (doc), pdf, and jpg images. So the fax machine itself isn't very promising. Its significance and the clue to the future is the self-configuring beeps. Suddenly a telephone was capable of graphic communication. Limited at first, but it was a harbinger of ultimate universal digital interoperability—machines talking to each other. If the machine is digital, it can talk to any other digital machine, which is an important clue indeed.

The earlier days of personal computers famously tested the patience of users attempting to connect peripherals like printers, cameras, or speakers to their computers. There may be the switching of switches, the finding of appropriate wires and plugs, typing in lists of settings and codes from awkwardly written manuals. The early Apple II computers introduced back in 1977 required users to actually solder connections, cut wires, use jumper cables, and manipulate tiny dip switches inside the computer to expand the computer's capacity for connections to printers, data storage, and display screens. As computers started to sprout an array of plugs on their periphery, an entire generation of geek squad specialists at big-box electronics stores made a career of getting good at figuring out which plug to use and setting up the correct "driver" software to permit connections. But that particular aspect of retail geek career expertise may be in jeopardy. Enter "plug and play."

This is the underlying logic that motivates our attention to the fax machine's chirping melodics. Plug and play usually works silently and invisibly. These days interconnected machines use Firewire (Apple's term for the IEEE 1394 standard) or USB cables. USB stands for Universal Serial Bus—the physical connectors that provide electrical power to and data to and from peripheral devices. To most observers, the familiar USB plug is just like the even more familiar plug-into-the-wall electrical plug. Just plug the male into the female. The key in this case, however, is self-configuration—the automatic mutual adjustment of settings of data speed and format, signal interrupt codes, input/output addresses, or direct memory access channels. When a USB3 (2017) or USB4 (2019) device connects with legacy USB devices, the software communicates and the newer device basically says "OK, old fella, I'll slow down my data speed and simplify my encoding so we can talk." The new standard is helpfully "backward compatible."

Because there are commercial incentives to keep customers brand loyal, not all plug-and-play connections across brand names work flawlessly. But user interest in self-configuring interoperability is also a powerful commercial incentive as we move into the future. Plug and play is only a few decades old. It has migrated from fax machines to potentially any machine. When two machines are physically connected by wire there is a requirement of a common standard of physical interoperability—compatible male and female forms of design. But as communication becomes increasingly wireless, even that potential impediment will diminish in importance.

Perhaps you are a bit skeptical of pondering the trifling technology of the fax handshake. No big deal, you say? The handshake itself, after all, doesn't have any influence on what is actually being communicated. It just sets up a few technical standards.

To make the case, let's take a few moments to consider a few of the many offspring of this progenitor of self-configuration. One of my favorites is the 1-Click purchase button for e-commerce. Perhaps you have encountered 1-Click on Amazon or similar e-commerce sites. You enter in your shipping and payment information just once and then simply hit the button to make any purchase thereafter. It has been an economic windfall for Amazon and, accordingly, perhaps a problem for the impulse-oriented shopper. Amazon had an idea it was on to something when it successfully patented the process in 1999. The patent continued to give Amazon an advantage over e-commerce competitors until 2017, when the patent expired. Seeing the broader significance of convenient self-configuration pays off. Apple saw the significance early on and set up a licensing arrangement with Amazon to use 1-Click for selling Apple products and music and video through iTunes. Barnes and Noble tried its own version of 1-Click early on without a license and ended up in court.

Now, many of us take 1-Click so much for granted that we may be more than a little annoyed at the prospect of typing in all the information when we are new to an e-commerce site. Not to worry, of course. Third-party software programs and apps will fill in the blanks for you. It is the auto-fill software industry. Perhaps you have run across RoboForm, LastPass, or Dashlane. Many services started simply helping you remember your passwords and expanded to names, addresses, birthdates, and credit card numbers. You may not have to seek out a service. Auto-fill may simply pop up as a preinstalled extension of the web browser you use.

Self-configuration isn't flashy intelligence, but it is intelligence. It is easy to take for granted. It is a puzzle piece in a bigger pattern.

It's All on the Internet—Where Does This Information Come From?

Imagine that a frazzled modern-day car buyer just returned from gladiatorial battle at the dealer's showroom. How many stories do you know about the clever tricks of car salesmen (it's usually the male of the species) and, most notoriously, used-car salesmen?

There is often some high drama in the showroom or on the lot. It isn't a fair contest. Dealers have every incentive to make the purchase of a car an emotional rather than rational experience for the customer. The salesperson's advantage is that they know the cost of the vehicle and every shiny option. Many options, it turns out, are priced at 500 percent of actual cost. Most buyers are vaguely aware that the manufacturer's suggested retail price (MSRP), also known as the "sticker price," should be a starting point for negotiation. But how would a customer have any idea how big the margins are that the dealer has to work with? What is the dealer's wiggle room? Good to know.

In the digital age, it would appear that the salesperson's advantage has been significantly reduced as the invoiced price-to-dealer cost for virtually every model is available online. A car buyer of even modest sophistication and experience can enter the showroom fully equipped. Not only do you know the dealer's invoice price, you know the so-called hidden holdback. It turns out that manufacturers send out "dealer incentives" so the dealers pay, in effect, less than the official invoice. Want to know the holdback? No problem. It's all online as well. (Ford is 3 percent of MSRP including options, Honda 2 percent of MSRP without options, Infiniti 1.5 percent of base, and so on.) Bring your smartphone. Use the dealer's free Wi-Fi to look up the details right in front of the salesperson. It would be the ultimate affront, except that most car sales are made by seasoned veterans no longer susceptible to such effrontery even at its tasteful best.

The thoughtful reader points out that, yes, MSRPs are negotiable, but supply and demand still comes into play. How do you know which ones are the hot models with waiting lists and transaction prices even above MSRP? Yes, again. It's all online. If dealer inventories are overstocked, on the other hand, the older models on the lot that have not moved are beginning to cost the dealer money and can often be purchased below invoice plus holdback. Much of that data is accessible from a variety of sources online. Repair statistics, expert reviews, and detailed individual vehicle repair histories are now available through independent commercial firms like CarFax. (Just for the record, it's CarFax rather than CarFacts. The name is a historical residue of the fact that the data was originally distributed by fax back in the 1980s and now, of course, by online databases.) Want to know if the dealer's deal on car financing is better than an outside firm or bank? The dealership's Wi-Fi is free. Look it up.

The underlying process for this second clue addresses the evolved and negotiated norms of *accessible information.* Economic theory dictates that a successful market transaction is the agreement between a willing buyer and willing seller. That works best when there are no informational asymmetries. The tradition for years at the car dealership was that dealers knew all about the cards they were holding and often a fair amount about the cards in the buyer's hand.

We focus here on the car purchase because it is a major economic decision and asymmetries in knowledge are particularly important. But the underlying issue of a seller's presumptive advantage in the marketplace is a revealing clue to the changing nature of economic exchange. Shall we mark up a clear win for the car buyer and a corresponding loss for the car salesperson and close the case? Perhaps not so fast.

A case in point is provided by one W. James Bragg of Long Beach, California, a consumer advocate and adviser on automotive economics. Some folks follow stocks, some follow sports; Bragg is a new car invoice data junkie. He got his MBA at Harvard and worked as an executive in retail and automotive advertising and then set out on his own as an author with several books on automotive pricing. His basic argument is that when invoice information went public on the Internet in the mid-1990s, the manufacturers and dealers made appropriate adjustments to protect dealer profit margins. He has abundant data about car sales below the officially circulated invoice prices (including holdbacks), which, of course, could not be sustained by any industry if the invoice data was actually accurate. Further, he shows us how historically invoice prices have been creeping up closer to sticker prices.

Bragg tells a wonderful story about confronting the senior executives of Consumer Reports, Kelley Blue Book, Edmunds, and other automotive advisory services at an "open and candid roundtable discussion" on "whether online car shopping and information services are believable and are relevant in today's market." This roundtable was organized by *USA Today*. He describes himself as the little David at a meeting of Goliaths in expensive suits. Perhaps a better narrative metaphor is the king with no clothes. He hands out a detailed report that demonstrates that the publicly circulated invoice prices all the big consumer information companies use are systematically inflated to hide actual dealer profits. He points out that in the 1980s, before the Internet, only 10–15 percent of car buyers had any knowledge of invoice prices, so their existence was not a big threat. In the increasingly digital 1990s, the car industry couldn't successfully make invoices disappear, so they systematically distorted the data. Bragg gives us something to ponder:

> Where do you think Consumer Reports and those auto-pricing websites get dealer invoice prices? From skywriters? Candy wrappers? Twitter? Cereal boxes? YouTube? Lottery tickets? Victoria's Secret catalogs? Believe it or not, that information comes straight from the horse's mouth—the automakers that set those prices.

Bragg says he handed out his report and the executives read it and you could hear a pin drop. Nobody disputed his data. The executives hemmed and hawed and muttered that their consumer research indicated that most consumers felt they got a fair deal. That may be true, but then:

> One participant said the exhibit showed that the industry had responded to the publishing of invoice pricing online by saying,

> "We need to find another way to mask what dealers are being paid so that people don't feel like they're getting screwed." No one disagreed.

That was a while ago. Nothing has changed since as far as I can tell. So what is going on here? Dealers continue to sell cars at a profit and convince customers that their deal is near or below invoice price. Consumer Reports and others continue to provide the market data they are given in turn to consumers. Bragg continues to sell books about how the invoice price system is fixed.

The moral of the story is that when economic and institutional incentives are strong, the data system is likely to be systematically biased. It is a familiar story in crime, medical, and some economic statistics. Pharmaceutical companies have many millions of dollars riding on drug trials. That leads frequently to the underreporting of failures and sometimes the exaggeration of success. Battlefield over-reports of enemy dead represent another classic case in point.

Norms of reliably accessible information, especially those relevant to important transactions, are an important element of the next stages of the information revolution. But the clue has a caveat. The digital age provides an abundance of information online—a messy mix of accurate and systematically and intentionally distorted information. Seek out the appropriate data on which to make decisions and proceed with great care. The data may be reliably accessible, but is the information reliably accurate? A clue with a caveat.

History's Most Famous Broken Laser Pointer—Engendering Trust among Strangers

On Labor Day weekend, Sunday, September 3, 1995, a 28-year-old Paris-born Iranian American named Pierre Omidyar launched an online service called AuctionWeb. Omidyar was a talented computer programmer who expanded his personal website to enable the listing of a direct person-to-person auction for collectible items. The first item sold on the site, it turns out, in what is now a well-known story, was a broken laser pointer.

As the legend goes, Omidyar felt his business presentations lacked pizzazz. His impulsive solution was to buy a $30 laser pointer to spruce up his PowerPoints. That would do the trick. History doesn't record whether he ever got to use it because it was soon broken. New batteries didn't help. So, what the hell, he listed it on his brand-new auction website—the ultimate test of his online auction concept. He was clear. The damn thing didn't work. He started the bidding off at a dollar and forgot about it. In the first week, there were no bidders. By the second week, someone had actually bid $3 and then $4. This was Omidyar's personal website, which originally focused mainly on issues related to the Ebola virus. There was no marketing. How did people find this most curious auction item? Particularly people interested in broken consumer electronics. By the end of the two-week auction, the bidding had reached $14.

Omidyar was astonished that anyone would want a nonfunctional laser pointer and contacted the winning bidder to ask if he understood that the laser pointer was broken. In the responding email, the buyer explained that he was, in fact, a collector of broken laser pointers. It foretells much of what is to come. Who had a clue there might be a market for such things? Other items that appeared on AuctionWeb those first few weeks as others experimented with the new site included a metal Superman lunch box ($22); Marky Mark Underwear, autographed, no less ($400); and a Yamaha Midnight Special Motorcycle ($1,350 asking price).

Similar surprises followed. The business exploded as participants began to list and trade goods of an unimaginable variety. What followed was a classic Silicon Valley success story. The business moved from Omidyar's spare bedroom to a tiny one-room office in a start-up incubator in Sunnyvale (managed, no less, by NASA), and then it moved again to a block of rooms in a nondescript business park across from a shopping mall in nearby Campbell, California. At first, the business had a few employees working at folding card tables, then a few software engineers turned Omidyar's code, cobbled together over a holiday weekend, into a working and scalable online auction site. By early 1998, the company had 30 employees, a half-million online participants, and revenues of about $5 million. In September of that year the company went public with an initial public offering (IPO) of $18 a share, which shot up 160 percent the first day, valuing the start-up at almost $2 billion. Omidyar and his fellow card-table-based employees were instant millionaires. eBay was to become a big online player dominating the auction and the related e-commerce business space, including creating and then spinning off transaction platform PayPal. Currently, the company has annual revenues of about $10 billion and 14,000 employees. OK, big success story. What is the lesson in this case? For the takeaway here, we need to return to the early days. For the first three years before the IPO, everybody was skeptical that this online auction business could actually work. How could it work?

Return to Omidyar's conversation with the company's very first potential employee, Jeff Skoll, who ultimately became the CEO. Skoll was very skeptical. Omidyar had met Skoll a few years before through mutual Silicon Valley friends and was impressed with his business sense and his freshly minted Stanford MBA. Omidyar recalls that it was around Thanksgiving 1995, when AuctionWeb was but a few months old: "I told Jeff there were people buying and selling on the Internet who never see each other but actually send money and stuff back and forth." To which Skoll responded: "That's ridiculous."

This is a telling exchange. Skoll was a sophisticated student of the evolving Internet business culture. He was working for the Knight Ridder newspaper chain overseeing its online experiments. Would online auctions work? When he first talked with Omidyar about the idea, Skoll had just returned from a conference on the then-emerging e-commerce concept. The conference's 300 attendees were polled on how many had actually bought or sold anything online. This was not a random sample. It was a group of high-tech e-commerce enthusiasts and start-up wannabes. Only three people indicated they had actually bought or sold something online. Not a promising start.

But a few months later, the following spring, Skoll changed his mind and left his secure job with a deep-pocket conglomerate to join up with Omidyar full-time and work initially in Omidyar's spare bedroom. What did Skoll see that others remained skeptical of?

There is a four-part answer. First, clearly the interest in this initially obscure site was taking off. Within four months of launch, there were thousands of auctions and tens of thousands of bids—all of that with virtually no marketing to publicize the site's existence. Second, the newspaper chain he was working for already recognized that online classified ads were cutting into the market share of old-school classified newspaper listings. Third, the online bidding process allowed for real-time dynamic pricing. Traditional classified ads typically required the seller to guess at the value of an object to be sold. Dynamic pricing dramatically empowers market forces of supply and demand to interact. And finally, fourth, somehow, someway, the all-important trust issue seemed to have been worked out.

The challenge was to introduce an element of trustworthiness into transactions between complete strangers. Each potential buyer and seller lacks an independent means of verifying whether a potential trading partner is reliable. Omidyar's solution, and what would prove to be key to Internet e-commerce culture, was a means of transparency—a *reputational feedback system.* Require that participants rate one another and make those ratings public as part of the auction process. It is not a perfect solution because buyers or sellers could try to game the system. They could invent multiple identities and always start fresh. They could do a few modest honest trades to build reputational equity and then pull some sort of stunt. But it has turned out to be good enough and is now pretty much ubiquitous in online trading systems. Despite some media attention to a few high-profile cases, fraud on eBay has been extremely rare. According to eBay's figures, less than 0.01 percent of its millions of auctions has resulted in fraud complaints. eBay, of course, has an extensive system in place for dispute resolution. The company has every incentive to keep the public's sense of trust in the process at a high level.

So our broken-laser-pointer millionaire has a lesson to share with us—a lesson critically important to collective networked intelligence. The lesson is the importance of reputational feedback systems. Back in the day, when most of us lived in tiny villages with a few shops and shopkeepers, everybody in the village knew the reputations of each. You know—make sure Big Bruno the Butcher doesn't put his thumb on the scale. He'll do that if you're not watching. In the modern world of near-universal digital connection of all beings, we have workable, scalable (although not infallible) means for keeping Big Bruno's thumb where it belongs.

Think of how often we rely on sources like Yelp.com for restaurant reviews, RottenTomatoes.com for movie reviews (both by plain folks and professional movie critics), Uber.com for driver ratings, and TripAdvisor.com for aggregated traveler reports on hotels and vacation spots. The list goes on. We take the technical capacity of real-time reputation systems for granted. Its particular significance lies in the fact that it is collaborative intelligence. When desired, an individual judgment can be inspired by a collective one. In reviewing the stages of human coevolution with evolving technology, we noted how a fully developed language allowed individuals to draw on the wisdom and experience of others. EI is a technologically enabled window on the world that affords a dramatic expansion of that uniquely human capacity.

Too Much Spam? Perhaps Intelligent Filtering Can Help

The BBC's Monty Python comedy crew picked the name Monty Python because it sounded silly. They built one of their most famous skits around the word "spam" for the same reason. The diners arrive at the Green Midget Café for breakfast and the waitress (that would be Terry Jones in drag, a rather predictable Python convention) recites the breakfast menu, which starts out unremarkably with eggs and bacon but soon deteriorates into increasingly ridiculous Spam concoctions. The final menu item, which Jones recites enthusiastically in his discordant falsetto, is "Spam, Spam, Spam, Spam, Spam, Spam, baked beans, Spam, Spam, Spam, and Spam." The café, of course, is inexplicably out of baked beans. This being a Python skit, a crew of Vikings soon arrive singing a Spam song. The skit was so popular that many variations cropped up in later shows and ultimately as part of a Broadway play.

Nobody is quite sure who first used "spam" to identify unwanted junk email, but clearly inspired by the Python skit, it caught on in online chat rooms in the 1980s and soon became an ever-present tag. Linguists point out the term's appropriateness given the near-universal perception that junk email is ubiquitous, unavoidable, and repetitive. By the late 1990s, spam was formally listed and defined in dictionaries, in both noun and verb forms, to mean unwanted emails. Hormel Foods, which produces Spam (the canned meat version of processed shoulder of pork and ham) and owns the trademark, tried to litigate against the name's unauthorized use, especially in commercial spam-fighting software products, but soon gave up. Hormel need not worry. Fans of the food product were undeterred, and Spam remains a successful niche food item.

Our concern here, and the telling clue to the digital future, involves the deeper underlying problem of abuse in open digital systems that the term spam identifies. Spam-filtering can be seen as an extension of the last problem we were pondering—the need for reputational feedback to distinguish legitimate from illegitimate information in the tidal flows of available online data. It costs next to nothing to send spam messages. So, if as a result of email spamming, many thousands of people are annoyed and one curious recipient responds, it still makes economic sense to continue spamming away—spam, spam, spam, and spam. Internet-initiated commercial junk phone calls are equally inexpensive, so the economics continue to motivate aggravating robocalls on landlines and cell phones, too.

Various Spam Kings have been fined in civil and criminal cases, and many have been incarcerated. One Robert Soloway, according to prosecutors, was responsible for tens of millions of unsolicited emails promoting his own company in the mid-2000s and was charged with identity theft, money laundering, and mail, wire, and email fraud. Authorities confiscated Soloway's 27 pairs of designer shoes and all his Armani and Prada jackets derived from his reportedly $20,000-a-day spamming proceeds. He served a little less than four years in prison and recently worked in a print shop for $10 an hour. Another heir to the title of Spam King who achieved infamy is a gentleman by the name of Sanford Wallace, who transitioned to sending online spam from a junk faxing business. His Cyberpromo Company was particularly adept at spam-blocking evasion tactics such as using false return addresses, rerouting through third parties, and employing a trick called multihoming. Wallace ultimately pleaded guilty to electronic mail fraud and criminal contempt of court and served just under two years in a federal prison in Colorado.

But as the economic incentives remain strong and overseas servers make the enforcement of anti-spam laws difficult, the legal task devolves into a game of whack-a-mole. Many more aspire to the Spam King title. Even the notorious "Nigerians" are still at it, promising millions of dollars in unclaimed funds if you'll just wire them a few hundred dollars to cover fees. It isn't limited to Nigeria. Many of these advance-fee cons originate around the world and have appealing names like the Spanish Prisoner scam, the black money scam, Fifo's Fraud, and the Detroit-Buffalo scam. It is hard to believe individuals continue to be suckered in and fail to recognize these familiar email appeals, which typically start out with the salutation Dear Friend in Christ or perhaps Dear Sir or Dear Beneficiary.

We focus on the obvious, perhaps even inevitable response to the problem that an abundance of spamming engenders—the automation of *intelligent filtering*.

Google's email software, for example, uses a complex algorithm to identify spam. It typically doesn't delete the likely spam. It tags the incoming messages and asks the users' help in weeding out unwanted messages. Google's statement says, "With Spam protection for Messages, we warn you of suspected spam and unsafe websites we've detected. If you see a suspected spam warning in Messages you can help us improve our spam models by letting us know if it's spam or not."

It can be a delicate trade-off between rejecting legitimate email versus accepting unwanted spam. Many email systems set up a spam folder that can be fun to skim through when other duties are less pressing. It is amazing how many plane crashes in Africa leave millionaire heirs who want your help in spending their new-found wealth, so many crashes that often several different heirs will appeal to you in the same day. And those consignment boxes, metal trunk boxes weighing approximately 110 kilograms each and apparently containing millions of dollars. They keep ending up at the Hartsfield-Jackson Atlanta International Airport with missing paperwork. Inspector James Edwards from the United Nations (not sure what he's doing in Atlanta) just doesn't know how to proceed. Can you help?

Spammers vary the names of those in need of your help, the locations of the plane crashes, the name of the lottery with unclaimed winnings, the dollar amounts in metal trunks, and of course, the source email addresses, but intelligent text analytic spam filters will recognize the patterns. The typical machine learning algorithm trains on a human-coded training set of mixed legitimate and spam messages and "learns," adjusting its set of rules from any mistaken false positives or false negatives. Spam filters even adjust to the personal styles of users. One classic case is a biochemist who actually works with Viagra field trials and would find that much of his legitimate work-related email ended up in a spam folder.

The spammers and scammers are smart and highly incentivized, so they typically analyze how the filters work and try to program around them. To stop text bots from entering discussion boards or engaging in fraudulent e-commerce websites, these systems ask questions that any real person but few bots can actually answer. Silly stuff like "What is the first letter of this paragraph?" Or captchas require decoding distorted text or graphic images. Sometimes the bots got smarter. Sometimes the bad guys simply hired low-cost human labor to try to get around the filters. Some spammers tried using images of text rather than the text itself to try to outsmart text analytic filters. Anti-spam programmers got good at recognizing the digital signatures of the image files. Cat and mouse. Spy versus spy. For years, 60–90 percent of email traffic was various forms of unsolicited spam. That ratio has been falling. And with the successful use of updated filters at both the sending and receiving ends, spam percentages in the active inbox of the typical email user have fallen to the low single digits. Intelligent filtering—pretty successful, all things considered.

Why is this spam business a worthwhile clue? It is because it addresses the inevitable challenge of an information economy—an overabundance of information. The spam example addresses information that is coming at you. Now look at it the other way around—when you seek out information. Here the exemplar of intelligent filtering is the all-powerful model of modernity, the search engine. Current estimates of the size of the open and accessible web are about 3 billion to 6 billion pages of information, and it's growing (almost doubling in size each year since 2012). That keeps Google and Chrome and Firefox and all the other search engines pretty busy. Do you usually find something pretty close to what you are looking for when you type a phrase into that search box (as it intelligently corrects your typos)? Yes? Pretty successful, all things considered.

Moving beyond the Tower of Babel—Instant Translation

Have you tried the Ambassador earbud? It uses the latest fast neural machine translation algorithms, machine learning systems that get better the more you use them. "Breaking the language barrier at home, at work, anywhere you need it," is how Waverly Labs advertises its automated translator. "Translate real-time conversations, menus and street signs while offline, websites, documents, and more using the Translator apps" on your cell phone. Just slip on the Ambassador earbud. It actively listens for someone speaking near you (within eight feet) and translates their speech into your native language in real time. Currently it translates 20 languages and 42 dialects—English, French, German, Italian, Portuguese, Spanish, Arabic, Greek, Russian, Hindi, Turkish, Polish, Mandarin, Japanese, Korean, Cantonese, Hebrew, Thai, Vietnamese, and Dutch. Sounds like this might be crazy expensive? As of this writing, the cost is $179. The real thing.

This amazing technical achievement may strike you as unremarkable. The last time as an English speaker you surfed your way to a website written entirely in French you simply took advantage of the inviting little button at the corner of the page that offered simply to "translate this page." You click the button, of course, and in a fraction of a second you're reading in English (although, given the current state of the art, the text probably includes a number of grammatical errors and linguistic approximations). Near-universal translation, it appears, is already routine.

Instant machine translation is perhaps the most revealing of all the clues at hand. Step back for a minute to think about what these devices are actually doing, the auditory versions in particular. The earpiece is listening to ambient sound and identifying a specific voice, recognizing the language spoken, accurately interpreting run-on phonemes as text, interpreting the text meaningfully (including elusive idioms), translating the text successfully, and converting text to speech—all in near real time. How do they do that?

The answer is an extraordinary amount of computational power and a whole lot of error-corrected practice. A lot. Researchers in this domain share a massive open-domain data set that contains over 25 billion parallel sentences in 103 languages.

OK, impressive. But why is this awe-inspiring computational exercise particularly telling in understanding the prospect of evolutionary intelligence? Well, because it empowers people to communicate successfully. Because the machine intelligence is relatively unobtrusive and easily taken for granted. Because these technologies are on a fast track and getting better at exponential rates.

In different ways, this is true of each of the elements of present-day intelligence we have examined. The underlying characteristics they share are:

Each empowers.

Each facilitates.

Each adds value.

Each updates with the latest information.

Each waits patiently to be called upon.

Each draws on collective experience.

Each learns from experience.

Each of these prototype systems in one way or another was designed to facilitate communication—the fax handshake, the informationally empowered car buyer, the reputational system that makes online commercial exchange between strangers work surprisingly well. Similarly, the hardworking email spam filter, the instant auto-translation from Arabic to English.

They are all technical, digital, and linked in various ways to the Internet. The fax, of course, preceded the Internet and was designed to expand the graphical power of the Internet's immediate predecessor—the global telephone system. And because the Internet does these sorts of things faster and better, online communication has largely relegated the fax machine to the storage room next to the 8-track tape player and the Walkman.

What is it about the transition from the telephone to the web that signals the common empowering character of our set of prototypes?

A car buyer could theoretically call up an appropriate expert and inquire about the invoice price. A laser-pointer collector could call up a stranger who advertised in the want ads and negotiate a good price for a sought-after laser-pointing rarity. And so on.

What does the Internet have that the traditional telephone system lacked?

Each of the communicative exchanges in this collection of clues could theoretically have been facilitated by the telephone. Sort of. But the web makes it so much easier, so much more practical. If we can understand the transition from the phone network to the Internet, we can better understand the transition from the Internet to EI. Where did the Internet come from anyway? If the United States Department of Defense had not built a digital network for its own purposes and then (rather absent-mindedly) given it away, would somebody else have pointed their finger in the air, shouting "Eureka, let's build a digital network"? Good question.

And here's the answer. Someone did. It was AT&T, which called its invention ISDN for Integrated Services Digital Network. You probably never heard of it. It went nowhere. And the reasons ISDN was a big flop help us understand what is unique about the Internet and why its existence is so critical to the character of our current digital era.

Western Union operated the dominant electronic communication system of the mid-nineteenth century: the telegraph. It was their monopoly. Western Union reputedly was offered the chance to acquire Alexander Graham Bell's 1876 patent for the telephone. It was the company's chance to dominate the next stage of electronic communication. As the oft-told story goes, the president of Western Union, one Mr. William Orton, declined the offer and said dismissively, "What use would this company make of an electric toy?" Western Union at the time had all the engineering talent to invent a telephone and had actually explored the technology, but it didn't happen. When the company recognized its error and came late to the market, it was outgunned by Bell's patents and withdrew from the growing telephone industry by 1879.

History repeats itself in the mid-1980s. AT&T begins to recognize the significance of a potential digital competitor and sets out with its impressive engineering talent (sound familiar?) to invent a digital telephone system. It's going to be revolutionary (in phone-company think). There will be not one but two digital voice channels. By golly, you could send a fax and speak with a colleague at the very same time. Each would have the fixed limited capacity of 64 kilobits. Why such a low bandwidth? Hey, it's perfectly adequate to digitally encode the human voice. What else did you have in mind? This is a telephone.

In the early 1990s, AT&T and many other national and international monopoly phone providers around the world tried to sell an ISDN upgrade at considerable expense with a dramatic lack of success as the Internet took off as a public network with a bright and much more expansive future. ISDN did get used occasionally for high-quality sound remote broadcasts for radio and for some early-stage videoconferencing—minor victories each.

The Internet, as it turns out, was largely an accident. A happy accident, but basically a series of design decisions that resulted in an unstoppable open, flexible, expansible digital global network. In retrospect, some of the early participants claim that they saw it coming, but there is room for skepticism about such early prognostication. The government team at DARPA (the Defense Advanced Research Projects Agency) was inventing a military data network for engineers and research specialists to exchange software. In early computer science, there was the tradition of a "comment card." Coders would put in a comment card (an IBM card, of course) that represented a line of text for humans to read that would be ignored by the computer; this text would allow subsequent programmers to understand what the initial programmer was trying to do at that point in the program. Comment cards started circulating online without the program. Email was born. It wasn't designed. It wasn't foretold. It just happened. The initial Internet protocol, the famous TCP/IP, wasn't patented. It was simply circulated publicly for comment. And the system had no capacity whatsoever for measuring usage and for billing users. It was a government data system. Like a government building, employees are not charged for space or for light or the electricity they use.

The reason the Department of Defense was building a packet-switched network in the first place was the vulnerability of the old circuit-switched telephone system. In a potential military conflict, the enemy could destroy a few telephone central offices and wipe out the capacity for strategic communication. Packet-switched networks are more agile and flexible. If the data encounters a broken or over-taxed router, it automatically switches to another route. There are no central-office weak points. And it is a data network. There is no fixation on voice communication and a needless limit of 64 kilobits.

By the late 1980s, the control of this new digital network had migrated from the Department of Defense to the National Science Foundation (NSF) and then to a series of private entities that were now free to commercialize it. The dot-com revolution was underway, not just in the United States but globally.

So the unanticipated magic key in digital design, motivated initially by fear of enemy attack, was built-in agility and flexibility. TCP/IP is designed to move packets. It doesn't process or interpret the data. When the phone companies offered data services to businesses, they wanted to add sophisticated specialized protocol processing to their services so they could charge more for it. One example of that was the X.25 data standard typically used to connect ATM terminals to central bank computers in the pre-Internet era. The Internet designers had no such motivation. They designed a "dumb" network so that intelligence could flourish at the edges. All the processing (the turning of ones and zeros into text, images, and video) and all the interactive video-game-like intelligence was in the terminal, the user's computer. The Internet just moves packets of data. Very fast. Very reliably. The Internet, unlike the phone company, doesn't want to add value and then bill you for it. That became the job of everybody else, and add value they did. The Internet never had a built-in system for billing users for use like the telephone system that was capable of keeping close track of every second of every call. It never occurred to the Internet designers. They were designing a system for the Defense Department. Soldiers get free rides in jeeps and trucks all the time. Never once was there a taxi-type meter to bill them. No one gave the issue any thought.

AT&T or any other similar entity in telecom, broadcasting, or computer networking was unlikely to design an open-standards packet network that, by definition, they couldn't control or exploit. So one might argue that it pretty much had to happen the way it did. It had to be an accident.

If AT&T (and the other major commercial network players of the 1980s) had realized what the Internet would become, they might have been successful in stopping it—arguing that a tax-based, government-sponsored system should not become a competitor in the marketplace. Or, if these players recognized the threat early enough, they could simply have bought up all the nascent players in the early Internet and, of course, redesigned it so that users would have to pay as much as the market would bear for each bit and every computation. When AOL (America Online) had a dominant position in connecting users through modems to the early Internet, it tried such a strategy. AOL called it the "walled garden." It tried every trick in the book to keep its subscribers dependent on AOL-controlled platforms. But it was a losing battle. We will return to the AOL case in the chapters ahead.

So, the key to the story is that the U.S. Department of Defense, with its interests focused elsewhere, casually gave away a quadrillion-dollar technology and let loose what became an unstoppable, universal, open-architecture digital platform.

The unstoppability is easy to understand. It is the well-known phenomenon of network externality—the value of the network is a function of the number of nodes (in this case people and institutions) connected to it. Any potentially competing networks by definition start with a limited number of participating nodes and are accordingly less functional in making potential connections. We have watched new connecting platforms start small and come to replace older ones. In the social media space, Friendster was replaced by Myspace, which in turn was overtaken by Facebook, for example. Successive platforms offered unique social dynamics and system functionalities so subscribers switched. But each of them was based on the Internet, because the Internet just moves the packets—the universal mover of bits. What the bits do is up to the "intelligence at the edges" of the network, the software on your laptop and the apps on your smartphone. Since the Internet itself isn't trying to control and derive profit for moving a particular type of bit—say, a very popular game or movie—it is innovation friendly. It is a truly open architecture. A digital free for all. The ultimate global commons.

A New Form of Human Connectivity

Universal Connectivity. My argument is that we got lucky. Unbelievably fortunate. Sometimes history smiles and winks at us. There were perhaps a half-dozen points in the last half century at which the Internet could have been stopped, redirected, or otherwise derailed. If one posits (and I believe it is reasonable to do so) that each fork in the road represented a 50–50 chance that things could have gone the wrong way, then we were very, very lucky. That's getting 50–50 odds right six times in a row. Multiply that out. It comes to about one chance in a hundred of ending up with the Internet as we know it. Think about what could easily have happened to derail it.

1. DARPA could have been ordered to redirect its research funding to another particular weapons program rather than computer connectivity. That sort of thing happens often enough.
2. The early developers could have depended on AT&T instead of building their own separate high-speed network. AT&T could have ended up controlling and even redesigning the system to look more like its pet ISDN project.
3. The early developers could have decided on their own that it was appropriate to protect the packet-switching protocols with patents.
4. As it turned out, teams of engineers all agreed on how to implement the standards. They didn't divide into competing camps with different incompatible standards, as happened with

television broadcast and many other computer communication standards.

5. The Department of Defense basically gave its system away to the NSF, which in turn gave it away to the world—no patents, no restrictions, no questions asked—with a full set of instructions on how to make it work.
6. Finally, the rest of the world chose to adopt the American-developed standards for what would become a global, interoperable digital network. That includes the Chinese, the Russians, the Iranians, everybody. How often does that sort of thing happen? The world got lucky.

Not only that, the system turned out to be remarkably scalable, robust, and extensible. It was originally designed to connect a few dozen military research centers to exchange text and computer code. The Internet has now been upgraded to facilitate connection to an unbelievable number of Internet addresses—3.4 times 10^{38}. That comes to 340,282,366,920,938,463,463,374,607,431,768,211,456 different individuals or devices. That should keep us going for a while. And it provides two-way digital connectivity between nodes at the speed of light carrying voice, text, image, or video with equal ease. Global, interoperable, high-speed, multimodal connectivity.

For the last two centuries of the Industrial Revolution, we had multiple unique and unconnected systems for human communication. Examples include the printing press, the postal system, the telegraph, the telephone, the motion picture, radio, television, the videocassette recorder (VCR), and cellular telephones. Each was technically optimized for a particular kind of communication. There was no interoperability. Before the fax, in order to send a letter by telephone, somebody would have to read it aloud into the mouthpiece. Ironically, that was what often happened in the very last days of the telegraph system. The Western Union operator would call the recipient and read the "telegram" over the phone.

The key to universal connectivity is the flexibility of the Internet's open architecture—a structural design not optimized for any preordained use. TCP/IP moves the bits. How they are organized and interpreted is open to the creative capacities of those who use the network.

Immediacy. You are at the car dealer. The salesperson has made an offer. Should you take it? "Just a minute, Mr. Salesperson, let me use your phone to call good ole Benjamin to see if he's home. He knows a lot about invoice prices and might clue me in." Good thinking, but not a practical strategy. Immediacy is key. It is a close relative of ubiquity. Computational intelligence is everywhere—embedded in the environment. Better than embedded, networked computational intelligence is the environment.

You are at a business meeting in a foreign country. The superior language skills of others at the meeting put them at a potential advantage over you. True, you can look up individual words in the Lithuanian–English dictionary in your hand, but it does slow you down a bit. Real-time language translation into your earpiece makes a big difference—it brings the prospect of a more level playing field. Immediacy is critical.

By immediacy, I am not just drawing attention to the dramatic speed of modern broadband connectivity. Immediacy means mobility. It is something that is always available, there when you need it. Immediacy is a product of two powerful technological trajectories—Moore's law and the growth of wireless connectivity.

Moore's law is the observation that the number of transistors we can fit on an integrated circuit doubles about every two years. Gordon Moore, the cofounder of Fairchild Semiconductor and later CEO of Intel, first made the observation in the mid-1960s. And this amazing trajectory has followed this course pretty accurately since then. Although the trend will slow down and ultimately stop in the next decade as it reaches the limits of miniaturization at atomic levels, the impact of this process is hard to overestimate. Because computer chips are manufactured from a photolithographic printing process, they are relatively inexpensive at scale. That means that computational power and data storage get more powerful per unit cost growing at a compound annual growth rate of about 40 percent. The dramatic character of Moore's law is largely lost on us today as we take for granted carrying a smartphone in our pocket with the computational power of what would require several room-sized computers costing many millions of dollars only a few decades ago.

When amazing computational power costs pennies, we find it everywhere: in light switches, light bulbs, thermostats—what has become labeled the Internet of Things. And that brings us to the related trend of increasing wireless connectivity.

Growth in wireless data rates approximates Moore's law, increasing by a factor of ten every five years or so. 5G (fifth generation) cellular technology, Wi-Fi 6, and Bluetooth 5 are rolling out. Each represents dramatic gains in the power and flexibility of wireless connection.

Cellular and Wi-Fi connect the individual to the networks. Bluetooth is the personal-level networking technology. It connects the smartphone to the smart watch and to the smart earpiece.

Networked Verifiability. Yes, networked. DARPA's unintended universal, flexible, unstoppable, digital global network adds new meaning to the capacity of computers to weigh evidence and come to a conclusion. In the prototypical example of stand-alone machine intelligence, the programmer provides the question to be decided and the evidence to be evaluated. The computer crunches the data and posits a conclusion. It is sort of a scary prospect both for the programmer and the computer. What are the search rules? What are the limits of a reasonable search? Weighing a dozen factors seems reasonable. How does one weigh a hundred thousand factors? They weigh the factors quite well, thank you. Consider Omidyar's eBay (and a variety of other market phenomena). Large-scale markets of hundreds of thousands of individual decisions work demonstrably fast, accurately, and dependably. In fact, the more ubiquitous the market, the better to avoid local biases and information gaps.

Verification is simply feedback in a working information system. Yours is the winning bid. You send $14. Omidyar receives $14. (Omidyar asks you if you know the damn thing doesn't work. You acknowledge, yep.) Omidyar sends you the laser pointer. You have received the pointer. You are thrilled and recommend AuctionWeb to others. That's feedback. (There is the prospect of false feedback and the injection of deliberately dishonest misinformation. We'll address that important caveat shortly.)

Self-Correction. If participants on eBay have a reputation for nonpayment or misrepresentation, others will be appropriately reluctant to do business with them. If a spammer has a reputation for, well, spamming, your email app may warn you, reroute the suspected email to a spam folder, or simply delete it. The spammers may try to obscure their identity, resulting in a cat-and-mouse game of keeping databases of bad guys up to date (the immediacy of verifiability). The capacity for learning from one's mistakes (and successes) online is critical and, it turns out, ultimately revealing. Early errors of autocorrecting and spell-checking in word processing and in awkward translation have become humorous clichés. Sometimes it was called the Cupertino effect—the word "cooperation," for example, was often autocorrected to Cupertino by older spell checkers with dictionaries containing only the hyphenated variant "co-operation." Prized autocorrect errors include the misspelling "definitely" as "defiantly," "Voldemort" as "Voltmeter," and the "Muttahida Qaumi Movement" being replaced with "Muttonhead Quail Movement" in a Reuters report. Then there is the placard at Chennai International Airport, where the Hindi is translated as

"Eating Carpet Strictly Prohibited." And Twitter features #GoogleTranslateFails, famously translating this colloquial Danish sentence, "The budget is a mess because Hans is not good with numbers, so, in short, he's screwed up the rest of the month," as "It's been the goat in the budget because His raining badly so quite short, he is on the bucket month out."

Fun stuff. Characteristically because colloquialisms get translated directly as rain, goats, and buckets. But in time, that's easy to correct. Young children learn to make sense of idiomatic speech from context. Translation algorithms can, too. There has to be some sort of feedback. And the feedback has to be reasonably accurate. And, of course, when ambiguity arises, translators can provide alternatives. As humans select which alternative works best, that is precisely the kind of feedback that improves accuracy. The more a language community uses a set of language algorithms, the more useful they become over time.

There can be problems with the feedback process, of course. What if the community using an algorithm is biased against, say, women or an ethnic minority? If the human feedback reflects bias, the algorithm following the feedback will as well, perhaps increasingly so as recursive processes run their course. Note that the human bias is the originating source of the problem here, not the prospective algorithmic process. The good news is that these processes (perhaps much more so than human and collective human judgment) are analyzable and, in turn, subject to discovery and correction. That becomes true when the algorithms are transparent and fully subject to correction. Famously proprietary and exceedingly complex algorithms like Google search are, by their nature, resistant to direct public scrutiny. But post hoc analyses and legal, regulatory, and public challenges given Google's prominence keep Google's internal audits active and attentive. The bottom line is that if Google's search results aren't providing what its users want, they'll click elsewhere—a decision at the collective level that represents billions of dollars. That motivates attentiveness. Google modifies the 200 factors in its search algorithm in small ways and in fundamental ones as well. In 2005, it began personalizing the process, so the same search for different people started to provide different results for the same text. A major change. Most modifications are minor—usually about 500 of those a year. There's no shortage of feedback data as users search and then click on one or more of the options—50,000 every second, three trillion a year. Feedback, anyone?

We have been puzzling over the puzzle pieces. Four, in particular—universal connectivity, immediacy, networked verifiability, and self-correction. When you put them together, what do you get? I posit that the answer is a remarkably new form of human connectivity, one that has the potential to change fundamentally what it means to be human.

7
Putting Evolutionary Intelligence to Work—Some Brief Case Studies

This book is an invitation for you to give some serious thought to how the dynamics of evolutionary intelligence (EI) will influence your life in particular ways. Given the differences among us in our personal, professional, and avocational worlds, EI may look very different. In all but a few instances, the earliest exemplars of networked intelligent devices have already made a big difference in how most of us live our daily lives. Perhaps you have wondered how you ever found an obscure address using an awkwardly folded map with excruciatingly tiny print now that Waze's perky GPS-based voice guides you there with thoughtful attention to current traffic patterns. Or maybe you are a veteran birder who has spent many a day in the field with trusty binoculars and a well-thumbed field guide. It turns out that birding is more than just sighting. Birdcalls are very distinctive. Now your field guide is digital and includes audio clips to help you make sense of what you've heard. Wouldn't it be useful if your field guide knew what rare species you need to fill in important holes in your life list? And would it not be especially useful if your field guide was connected to others in real time to signal you when that elusive Kirtland's warbler was seen by another birder nearby? Would some old-school birders scoff at such new technology taking the sport out of the sport? Perhaps. Go back enough years and some old-school birders might have scoffed at the very idea of binoculars.

We will try to dig into some case studies in this chapter to explore how EI may interact in unique ways with particular domains of life. Four domains—medicine, law, employment and finance—have technical and data-oriented characteristics that make them likely cases, so they were selected accordingly. Two others that would seem to be more of an odd match—romance and amateur athletics—were picked as well because of their diversity and distance from the typical world of computational decision making.

Personalized Medicine

How will EI interact with the future practice of medicine? Two themes emerge here. The first deals with a delicate balance between too much data and too little. The second deals with the limits of the human capacity for self-control.

First is the data-balance issue. Most of us who lack extensive medical training still have a basic idea of how the human body works. We understand that the body is an extremely complex data processor drawing on many thousands of self-contained electrical and chemical signals at any moment. There is, for example, the body's evolutionary capacity to adjust oxygen and sugar levels in the blood and the lymphatic system's response when threatened with infection. All of this is subliminal. You may be only vaguely aware of this systemic bodily communication unless the signals involve pain, hunger, sexual stimulation, or the need for rest, because those require a behavioral response. An individual seriously ill in the hospital is routinely connected to as many as a dozen external monitors to measure breathing, blood pressure, blood chemistry, heart rate, and the like. External monitors are required because internal ones may have failed in some way. And, given what might be characterized as the relatively primitive current level of medical practice, these monitors are bulky and incapacitating.

In the future, virtually all of these many thousands of chemical and electrical signals will be monitored unobtrusively. Think a Fitbit or an Apple Watch on steroids. Think *Star Trek*'s handheld medical tricorder. When perfected over time, sophisticated sensors will signal otherwise undetected danger signals of a potential seizure, stroke, heart attack, or similar event and call for preventive measures or medication. As these systems mature, the challenge will be getting the balance right. One could imagine an otherwise healthy young individual interrupted multiple times a day with a variety of false alarms. Will we pause six times between lunch and dinner to take yet another prophylactic pill? The issue, as is often the case, will be evolving an appropriate balance.

There will be a parallel issue in the future with our unobtrusive but omnipresent little medical tricorders. These technologies will not only monitor our bodies; they will monitor our environment. You may recall from the opening pages of this book that when Arnold Schwarzenegger's Terminator character confronted a biker at the bar, the soon-to-be-dispatched biker blew cigar smoke in his face. The digital readout in the Terminator's line of sight noted: "Threat assessment: Scan Carcinogen Vapor." One can imagine a future full of accessible sensors indicating environmental threats of varying levels of significance—high levels of ultraviolet light, bacterial or viral concentrations, pollen levels, and particulate pollution. Some of these indicators for broad geographic areas may already be accessible on your smartphone. A germaphobe or hypochondriac might never leave their bedroom. The good news is that evolutionary intelligence means that thresholds will be set to your personal preference and your personal sensitivities.

Because medical decision making so often engages questions of life and death and almost always the quality of life, medicine has been an early focal point for the refined application of data-driven intelligence. The surgeon, for example, is likely to have multiple monitoring devices tracking bodily functions and augmenting visual information in real time. Signals to the doctor may be visual, auditory, and tactile. Consulting medical specialists may be observing an operation by means of high-resolution video in real time and giving advice. Obscure but relevant medical records may be consulted, as appropriate, in real time. Given the special character of medicine, for those of us who work in other domains, we can usefully study the medical case as a harbinger of things to come.

Now for the second issue of self-control. We all understand that our behavior has a major influence on our health. This raises an especially interesting question. We are fully aware that a double cheese and bacon burger is a caloric hazard. Do we really want a nagging signal reminding us each time we succumb? When we are subject to addictive behaviors like smoking, would an advisory signal make a meaningful difference to our behavior? My friends in high school who smoked took special pride in the bravado of calling cigarettes "cancer sticks." The riskiness of cigarettes may have been part of their appeal at that point in their lives. Statistics reveal that each cigarette will likely shorten your life by ten minutes. The life expectancy of a smoker is typically eight years less than an otherwise comparable but nonsmoking individual. Such actuarial calculations are generally well known and fully understood by smokers and nonsmokers alike, but typically the data are casually if uncomfortably ignored. Whether EI will be able to persuasively formulate personalized information to influence real-time behavior is an open question. The good news is that public acceptance of this science has led to overall dramatic declines in smoking—a 67 percent decline since 1965, for example, in the United States. My guess is that cancer researchers will explore a variety of techniques probably starting with creative smartphone apps along these lines. For those who have absolutely no interest in quitting, there may be little that could make a difference. That is the advisory nature of EI technologies. Alarm bells ring and we ignore them. Then, in weary annoyance, we turn off the alarm or in anger throw the noisy alarm out the window. It is unlikely that that part of the human condition will change much.

Intelligent Romance, Really?

How you will react to this section of this book will depend on how old you are. The older you are, the more skeptical you may be about the value of computers and networks. The younger you are, well, online and romance, of course, duh!

I thought it might be useful to try to tackle this topic because the delicate and unpredictable chemistry of human relations, especially romantic relations, seems the farthest thing from the capacities of smart computer apps and network interfaces.

It turns out it isn't the farthest thing. When it comes to finding prospective partners, online network dynamics work pretty well. It is an application of sociologist Mark Granovetter's famous insight about the strength of weak ties. Our close ties, those close friends and family, tend to know about mostly the same people (and potential partners) we already know. By exploring contacts among our more distant "weak" ties, our horizon is broadened with many more promising candidates. Online dating platforms extend the horizon even further. Sociologists call this *disintermediation*—take matchmaker Aunt Millie out of the loop as the human intermediary and substitute an online platform. For many years couples met in school, in church, or through family acquaintances. Now the clear and undisputed primary romantic medium is meeting online.

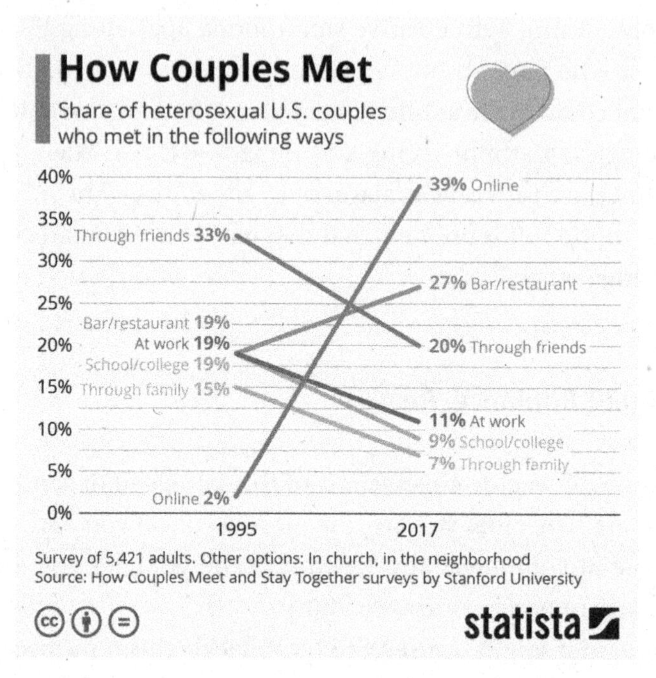

"How Couples Met" is licensed under CC BY-ND 3.0. *Source*: Katharina Buchholz, "How Couples Met," *Statista.com*, February 13, 2020, https://www.statista.com/chart/20822/way-of-meeting-partner-heterosexual-us-couples/.

A few questions arise right away. Don't people lie when they fill out those dating questionnaires? Yes, but only a little bit. They hedge their age by a year or two, their weight by a few pounds, and their income creeps up a few thousand dollars. Since the real-life personal relationship will soon reveal the truth, wholesale fabrication isn't going to work. And most entirely fraudulent dating applicants are detected by the platforms. One such platform promises to exclude 85 percent fraudsters within four hours.

Do the personality and interest-matching algorithms work? They aren't foolproof, but they probably work as well as Aunt Millie in judging compatibility. Since our self-description of our personality may have obvious biases, some dating platforms include psychographics derived from posted social media texts. The jury is still out on the value of that.

Part of the growth of web-mediated romance has resulted from the move of network access from the desktop to the smartphone. Grindr, the popular app for gay and bisexual men, pioneered geospatial location information. Users could not only see pictures of other active members, but the app also describes their geographic distance in real time. Humans, of course, have all sorts of verbal and gestural signals for flirting or demurring or to hint interest in future joint activities. It is a reasonable guess that the future of EI will involve a mix of physical and electronic signal exchanges among potential partners. It is intriguing to imagine how that would work. My guess is that more than a few science fiction authors have explored this terrain.

What about long-term relationships? One of EI's most promising potentials is the capacity to facilitate negotiation. There is the prospect of an omnipresent Digital Couples Counselor. I'm not sure how that would work. I am picturing the husband being advised not to say something known to drive his partner to distraction. I'm picturing the EI systems of each member of a partnership advising each to take a time-out. In a married couple's drawn-out battle over their child's best option for schooling, say, public versus private, I could envision EI systems suggesting other alternatives that neither would have considered. It is a promising domain for speculation.

Digital Law, Of Course

The practice of law involves adjudication, negotiation, the application of rules, and formal procedure. It would seem to be a domain well suited to EI applications.

Legal advice can be expensive, so one way to look at the prospects in this domain is to make a version of routine legal counseling available to a wider community at lower cost. One observer estimated that 80 percent of current legal work is "standardized" and that such standardized work may become fully automated within a few decades. Many elements of relevant legal advice are primarily factual. What is the speed limit here? Is that tax deductible? What is the applicable statute of limitations? Much legal advice is strategic. If I deduct that expense, what are the chances I'll get caught or increase the odds of getting audited? Is formal incorporation in the state of Delaware for my small consulting business worth the investment?

Many legal questions are subtle and nuanced—the sorts of questions that distinguish a very good lawyer from a mediocre one. It would seem at some point that an EI system would conclude and exclaim in effect: "My legal resources are limited. Best to consult an appropriately specialized human lawyer on this one." Many current online legal resources are designed to put the individual in touch with a well-matched human legal adviser rather than substitute for one. Evolutionary intelligence as described here is not simply a free-standing artificially intelligent agent or robot. It is a communication gateway to an online network of resources that, as appropriate, includes actively interacting human beings. We are reminded of the classic Turing test that was conceived at the dawn of machine intelligence. Are we interacting with a person or a machine? In the future, the distinction will be ambiguous and blurred, at times bit of both. In the future we may not notice or even care. In Australia, AskSomeone.Org.AU offers a 24/7 human-based online service for individuals confronting or fearing domestic violence and provides counseling and legal services in real time. That strikes me as an excellent resource and a harbinger of the future in multiple spheres of our lives. It also raises the issue of how EI will engage with evolving criminal justice systems. It is an important topic, certainly worth a book-length exploration in its own right, but out of range for our current exploration.

On relatively simple questions like wills and straightforward contracts, it is already the case that pro forma legal templates are potentially useful and available from sources like LegalZoom and Rocket Lawyer. There are multiple apps and websites designed to explain, presumably in real time, one's rights and obligations when stopped by law enforcement. Paralegal services such as notarization and copyright registration could easily be shifted to the digital domain and, in many cases, already have been.

Analog Athletics and Digital Intelligence

Amateur athletics is another candidate for a domain of human activity among the least likely to have much to do with machine intelligence. Yes? Let's take a look.

Want to get into shape with a new routine of diet and exercise? How about an in-person personal trainer? In the United States, personal trainers charge from $35 to $120 per hour. The presence of another human being, especially a knowledgeable one, with a collection of tried-and-true techniques and a carefully honed mix of soft empathy and hard discipline is what you need. No machine substitution here?

For starters, there is Peloton Inc., the franchise of high-end exercise bikes and treadmills equipped with video screens that provide real-time encouragement from live or recorded instructors, often with other spinners electronically present. Perhaps you have limited room in your cramped apartment for exercise equipment. What you need is a new video wall mirror called The Mirror. It looks like a normal wall mirror when not in use but comes packaged with resistance bands and a heart rate monitor and transforms into a connected fitness screen during workouts. You have access to live and on-demand workouts, including cardio, kettlebell, resistance-band training, yoga, boxing, and Pilates. The mixture of instructor demonstrations and your actual visual reflection mimic the role of the traditional trainer or sports coach. As intelligent image analysis improves (not there yet), the artificial intelligence (AI) coach will tell you to straighten your left leg a bit, yes, that's it. Currently these systems routinely measure your heart rate. In the relatively near future, the capacity to assess a much greater variety of bodily functions such as hormonal balances will be routine.

Professional athletics draws very sophisticated advisers. A qualified caddy for a professional golfer is likely to earn $100,000 a year. That level of salary is not in return for just lugging golf clubs around the course. These caddies are sophisticated advisers who also study the weather conditions, the course, and the competition. I'm not a golfer, but it strikes me that in addition to advice on your swing from everybody else in your foursome, in the near future you will also be digitally advised. Perhaps, finally, some improvement in your handicap.

The prospect of EI in the domain of amateur (and also professional) sports and competition raises another most interesting prospect: When is such intelligence cheating? The rule of thumb is that your coach can coach you all they want during practice but not in the actual match. In football and basketball, a coach can send in a play, but in most cases the competitors need to make decisions on their own during competition. No performance-enhancing drugs. No performance-enhancing advice. Electronic advice giving will be very hard to detect. It's going to be interesting. Stay tuned.

Cyberfinance, Not What You Think

There are more than a few computers whirring away on Wall Street. We can all conjure up images of the brokerage trading floor with as many as ten computer monitors flashing the latest data at each and every desk. Of all the dominions of human endeavor, few would exceed high finance in the degree to which computational intelligence has permeated the field to its very core. "Fintech" is the buzzword for financial technology that has become popular as computational intelligence has migrated out from the already computerized back-end systems of established financial institutions. It's "not what you think" if you are thinking of the stereotype of Morgan Stanley in New York routinely sending a computerized billion dollars to Deutsche Bank in Frankfurt over the SWIFT network. Those are the big guys. EI signals the migration of computer-assisted investment from just the big guys to the little guys—the casual investors and what we might label retail finance. Previously, in the discussion of networks and romance, we talked about disintermediation, which in that case was about removing an intermediary like your Aunt Millie in finding a love match. In this case, disintermediation means proceeding without your friendly local banker and broker.

One of the more fascinating political kerfuffles in the United States in recent years was the skirmish over the Department of Labor's proposed fiduciary interest rules during the Obama administration. The proposed conflict-of-interest component was designed to close legal loopholes permitting financial and retirement advisers to recommend investment products that are especially profitable to the adviser and not necessarily in the best interests of their clients. That should hardly be controversial, but it was. A proposal to formalize this ruling in law was defeated. Financial advisers in the United States are not legally required to serve the fiduciary interest of those they advise.

Recall our extended analysis in chapter 6 of car dealers, who traditionally had all the information (and experience), and the relatively naive car buyer, who did not. This asymmetry in the marketplace is so commonplace and important that Nobel prizes have been handed out to economists who have documented the dynamics. (That would be George Akerlof, who won the 2001 Nobel Memorial Prize in Economic Sciences for his work on what he famously dubbed the market for lemons.) A persistent problem in applying the ideals of economic theory to the real world is that those individuals and institutions closest to the actual gears of the economy figure out techniques that give them advantage and use political leverage in various ways to protect that advantage. Digital disintermediation potentially removes or changes the roles of brokers and advisers. The capacity for casual investors to buy and sell stocks without paying intimidating brokerage commissions has become commonplace. There are currently over 100 robo-advisory services bringing sophisticated investment algorithms to a broader audience at lower cost when compared with traditional human advisers—a breakthrough in what were once exclusive wealth management services.

Employment in the Digital Age

It is natural enough for workers to fear machines that may take their jobs. There is a long list of colorful stories from the industrial era in Europe as workers in fits of fear and frustration attempted to destroy the offending machinery.

- Anton Möller of Prussia invented the ribbon loom in 1586. He took his invention to the Danzig city council for a patent. The patent was refused, and the council put out an order to have him strangled. Fortunately, nobody followed through. He died in 1611 of natural causes.
- We don't want to forget James Hargreaves of Lancashire, England, who invented the spinning jenny. He made the first for his own personal use but then made a few more and started to sell them until 1768, when a group of spinners broke into his home and destroyed them all. When he tried to set up a factory nearby in Nottingham, his new establishment was set upon by another mob.

- In a movement peaking in 1812, frustrated weavers began smashing machinery and burning mills reportedly at a rate of 200 a month. They claimed to be led by a "General Ludd," which provides us today with the term Luddite. After months of continuing violence, 14,000 soldiers were dispatched and the "Destruction of Stocking Frames, etc. Act" declared machine smashing to be punishable by death. Tempers cooled.
- The pattern continued well into the nineteenth century when, for example, a group of protesters called the "Plug Rioters" in Yorkshire in 1842 would pull the plugs out of steam-engine boilers because they felt the engines were lowering the value of their work.

These were primarily symbolic protests. And sympathetic historians today claim that most of those involved were not necessarily anti-technology. They were just trying to negotiate fair wages for continuing work as the factories became more productive and profitable.

I'm not aware of any modern movement that makes a habit of smashing computers, but the serious concern about technological unemployment is widespread and prominently articulated.

In the middle of the Great Depression, economist John Maynard Keynes warned of "unemployment due to our discovery of means of economizing the use of labor outrunning the pace at which we can find new uses for labor." Wassily Leontief picked up this theme in the 1980s, publishing a famous article entitled "Is Technological Unemployment Inevitable?" But as Daniel Susskind explains:

> People have periodically suffered from bouts of intense panic about being replaced by machines. Yet those fears, time and again, have turned out to be misplaced. Despite a relentless flow of technological advances over the years, there has always been enough demand for the work of human beings to avoid the emergence of large pools of permanently displaced people.

Will that continue to be true as artificially intelligent systems begin to pervade the intellectual work of the white-collar occupations and not just the routine mechanical work of the factory floor?

Short answer: Yes. And that yes resonates with the central themes of this book as we have emphasized the potential complementarity of computational intelligence rather than its substitution for or competition with human intelligence.

Take, for example, this case study in medical diagnosis drawn from Susskind's research:

> In 2016, a team of researchers at MIT developed a system that can detect whether or not a breast biopsy is cancerous with 92.5 percent accuracy. Human pathologists, by comparison, were able to achieve a rate of 96.6 percent—but when they made their diagnoses with the MIT system alongside them, they were able to boost that rate to 99.5 percent, near perfection. The new technology made these doctors even better at the task of identifying cancers.

One of the most widely cited exemplars of computational intelligence's promise is IBM's Deep Blue. In 1996, the Deep Blue computer and specialized chess-playing software defeated then world chess champion Garry Kasparov in an official six-game tournament in Philadelphia. But consider what happened next. Kasparov began to champion a new approach that he called Centaur Chess. (The centaur, of course, is a legendary half-human, half-horse creature.) The idea is that a human working in tandem with a chess-playing computational system could beat any computer-only competitor. It builds on the intuition, creativity, and empathy of the human with the computer's brute-force ability to evaluate an incredibly large number of chess moves, countermoves, and outcomes.

There is perhaps a deeper lesson here. Technology historians have noted multiple "AI winters" in the second half of the twentieth century as progress in computational intelligence hit roadblocks, confidence and funding declined, and many researchers moved to other areas. Recall that the original underlying model for the field of artificial intelligence was basically to reproduce human intelligence—to impart human knowledge into computers as mathematical and algorithmic "expert systems" by modeling what humans do.

The most recent and most dramatically successful efforts of the last two decades in AI technology, however, follow a somewhat different path. Although the common vocabulary of deep learning and neural nets sounds human-like, its multilayered mathematical modeling uses millions of learning cycles that independently develop attribute models and may indeed have millions of dimensions and that work well but are simply incomprehensible for humans. In other words, computers, given the chance, think differently—leveraging different intellectual capacities that can complement human intelligence.

As noted at the outset of this book, our enterprise is serious. Important issues of equity, privacy, fairness, and social organization are to be addressed. And it is great fun to speculate about the future. Clearly, our attempts to extend current trends forward several decades will result in a mix of hits and misses. I hope to get more than a few guesses right. But the primary goal is identifying the important underlying issues and stimulating thought about how these technologies will be designed and controlled. The flavor has been largely positive and optimistic. These technologies offer great promise. But there is a deeply embedded challenge, a dark side only hinted at so far. This is the focus of the next and final chapter.

8
Next Steps

The purpose of this book is to provide the reader with a new perspective on a cluster of new technologies that lay ahead. It is great fun to speculate about how these new gizmos will work. But there are also serious questions about potential social, cultural, and economic consequences which, in turn, merit serious thought. We have employed the rather expansive phrase evolutionary intelligence (EI) to characterize these developments in an effort to signal their looming significance.

At several points in our review, we have described these technical trends not only as possible developments but rather as inevitable ones. It is important to be clear about what is being asserted. The growing ubiquity of computational intelligence in our environment is evident and hardly controversial. We already routinely use smart appliances, drive smart cars, and converse with smart speakers. The challenge is to address what happens when the capacity for intelligence moves from the exterior device closer and closer to ourselves and becomes intertwined with our perception and communication.

Other analysts have addressed this in other ways. Ray Kurzweil, for example, has predicted a convergence of biology and technology into a singularity—the bionic human. In such a case it might be difficult to distinguish where the machine ends and the human begins. I'm not sure what to think of such a proposition. It is unsettling, to say the least. This book confronts a set of developments that are likely to happen sooner and may influence or even preclude such a later development. EI is an immanent stage in human development when digital communication tools become very flexible, mobile, agile, convenient, resourceful, helpful, personalized, and ubiquitous. They may become so useful that we do not need to worry about surgically implanting them in our bodies and brains.

This book is one of a number that peer into the future. Our perspective, however, has been largely hopeful and optimistic. That may put us in a distinct minority among the more typically dystopic members of the prognosticator set. There are seven reasons for our optimism. They represent a number of our central themes.

First, these technologies offer a truly unique capacity to compensate for the well-understood weaknesses and biases of the evolved human cognitive system. We humans tend to jump to conclusions, pay too much attention to distracting details, forget things, misinterpret the motives of others, succumb to wishful thinking, and stubbornly stick to old ways. Because of their ubiquity and increasing contextual awareness, intelligent technologies can advise, remind, alert, warn, predict, correct, suggest, and point out to us conditions and situations as we make our way through the world. Most of us take some pride in making our own decisions. We may choose to accept advice but not direction. Advice may be provided, but only when welcomed. It is likely that individuals and their digital resources will develop a rhythm of when and where to engage. And we are indeed free to ignore advisories entirely. In turn, a similar rhythm may evolve when we learn over time that ignoring some forms of sound advice comes at our peril. We learn when to be especially attentive. Eyeglasses and hearing aids enhance our ability to see and hear. EI enhances our capacity to think. No one is predicting a world populated entirely by Einsteins. But even a minor improvement could go a long way.

Second, these technologies are typically easily manufactured. They will be inexpensive and ubiquitous. Circuits, once designed, are simply printed on a substrate and become very inexpensive per unit at scale. The current printing process is routine photolithography, and the current substrate is silicon (i.e., purified sand). The circuits are typically general purpose, so the software can be updated and the applications can learn and improve from experience without requiring new hardware. As a result, the benefits of these technologies can be widely shared and function to diminish inequities in the distribution of and the derived benefits from advanced technologies.

We may imagine that special skills or training are required to take advantage of high technology. Would you need a PhD in computer science to get these things to work? The underlying idea of evolutionary intelligence is just the opposite. True, the first computers were designed to be used by software engineers. We got better and better at human–computer interfaces. It took four decades to figure out how to use a mouse controller and windows graphic displays to make a personal computer practical. As our interfaces improve, the skill-based and effort-based digital divides decline.

Third, these technologies are not stand-alone devices. They are networked. The first networking technology was the invention of language. It allowed us to convey valuable information across distance and across time. Michael Tomasello, as noted, makes the case that this unique human capacity within the animal kingdom to share insights and observations is what makes our accomplishments possible. So, unlike the stand-alone wheel or engine or adding machine, these new technologies are intelligent nodes on an immense network. The network is, in fact, global. We have emphasized how fortunate this unexpected development turned out to be. A global network of instantaneous communication allows for the free flow of new knowledge. As we are all well aware, with the universal capacity of all 8 billion of us to speak as well as to listen, the global cacophony includes a messy mix of information and misinformation. Fortunately, increasingly intelligent and agile filters help to clear the wheat from the chaff.

Fourth, one of the defining characteristics of the next generation of software systems is self-correction. These technologies learn and improve as they work. The successful learning process, of course, requires feedback. And the feedback needs to be accurate and unbiased. Here our optimism hinges on the first point above—the capacity to adjust for the innate biases in human perception. Think about the capacity of humans to become "smarter" as they grow older. We learn from experience. The more experience the better. Except when we don't learn because we ignore and misperceive. We anticipate a coevolution of technology and human behavior as both get better and better at accurately interpreting feedback from their environment.

Fifth, these technologies permit personalization. This process draws on the dynamics of learning and self-correction as noted above. The technologies become attuned and aligned with the goals and aspirations of their users. One size doesn't fit all. We may aspire to achievement in very different domains. One among us focuses intently on athletics, another on stamp collecting, another on a legal career, and another on environmental sustainability. Looking at the world through my prescription glasses is likely to be disorienting and out of focus for you. But they work for me.

Sixth, intelligent global connectivity will not inevitably deprive us of our privacy. Just the reverse. It has the potential to give us control. In the current economic model of connectivity through mass media and social media, the value of our attention and the value of our personal information accrues to the owners of the media who, in turn, sell it to marketers. As it is famously said, we thoughtlessly give away our valuable personal information in order to view a few funny cat videos. If personal information can be monetized and the value can accrue to the individual, entrepreneurs will figure out how to make that work. The more entrepreneurs competing to make that possible, the larger the potential share of that monetization to the individual as they compete for your cooperation. Thus, for those that value privacy the most, the personal information is simply not for sale. For those unconcerned or for those who are happy to trade access to personal information for convenience—it's a sliding scale.

Seventh, we need not fret about the prospect that the computers will take complete control. Science fiction has provided us with an ample supply of killer robots and self-motivated HAL computers. It is the stuff of very engaging narratives. But it is not an unsolvable problem. Our characterization of evolutionary intelligence emphasizes the principle of human agency. The human behaves. The software provides informational resources. The alignment of human goals with computational capacity is the essence of how such an architecture is envisioned.

As we review these encouraging themes and observations that have been interwoven through the chapters of the book, it comes to mind again that we are reviewing technological affordances not technological effects. The dynamics of modern political economy, if recent history is to be a guide, may steer us in other directions. We make note the potential for the free flow of information through the global network. There are many vested interests strongly opposed to that prospect. We propose that you can monetize your personal information or, if inclined, choose not to monetize it. There are powerful forces that would prefer to do the monetizing for you. We predict these technologies will be broadly available at very reasonable cost. The dynamics of capitalism don't always work out that way. So we conclude on a note of caution.

Typically, books like this one end with an emphasis on the upbeat and words of inspiration. (Cue the uplifting theme music.)

But there is a reason for caution at this stage. Note that this chapter is titled "Next Steps." Our focal point thus far has been the technological capacities that we will confront several decades hence. This chapter steps back and asks, How do we get there from here? What do we need to do now?

So far we have been telling a story that has lacked a clear-cut bad guy. The good guy is evolutionary intelligence—a projected capacity for individuals to exploit the next generation of mobile and personalized computational technology to significant individual and collective advantage. But an engaging story usually requires a bête noire, the classic bad guy. The protagonist of the narrative requires an antagonist or two in order to test his or her honor, strength, and persistence.

We reviewed a number of implementation challenges in chapter 4. Among them were privacy issues and the possible atrophy of human skills. But they were largely judged to be manageable problems, worth careful attention, but not fundamental threats.

The problem isn't the technology. The idea that a super smart computer system could outwit us merits attention. But it is a potential challenge in the much, much longer term and, in my judgment, ultimately represents an issue with perhaps a half-dozen promising solutions. No, the immediate problem is entirely human in origin. (Perhaps you saw that coming.)

The most significant problem in my estimation is the prospect of system capture—a coordinated attempt to corrupt what would otherwise be an open system and to bias the typical EI interface among humans to sow discord and to accrue special advantages for certain groups.

Readers will recognize that this is indeed an old and familiar issue. Figure out a way to twist the marketplace to your advantage. Cartels, trusts, quasi-monopolies, and price-fixing deals go back to the spice trade on the ancient Silk Road, if not before. The cases I am most familiar with are the railroad networks of the nineteenth century and the broadcast networks of the twentieth century. These historical models are useful tools for understanding the future because the power dynamics then as now hinge on the critical question of network access—the enduring and highly motivated search for an economically relevant market choke point or bottleneck. The EI threshold is a promising candidate.

The mainline railroad industry that dominated transportation economics in the late nineteenth century was reasonably competitive. Shippers in Chicago could cut deals with any of four competing through lines to the East Coast. But the rural farmer did not have direct access to the main lines. The farmer had to deal with the local rail spur, a local rail monopolist who often would charge exorbitant fees. The deep collective frustration of the famers with local rail tycoons was largely responsible for the initial antitrust legislation in the United States—the Sherman and Clayton Acts (which in time also addressed the evolving oil trusts as well). The same choke-point dynamics dominated the network television industry in the mid-twentieth century. The independently produced television programs in Hollywood had no way of reaching an audience without negotiating with the New York–based networks whose 200+ local affiliate stations had the only broadcast licenses available. (In the previous generation, the Hollywood studios had bought up most of the movie theater chains so they could control the theatrical distribution network until forced to sell by an antitrust settlement in 1949.) Railroads and television networks were remarkably profitable enterprises in their days of dominance. If you control network access and can regulate the individual's capacity to get goods or information into the open network, you have profit-reaping power. One observer called it the next best thing to a license to print money. This sort of thing attracts attention.

Imagine the precursors of an EI interface in the immediate future. Perhaps you're hungry; you rev up the car and program your GPS system for the nearest Applebee's. But on arrival, you find yourself in the Olive Garden parking lot. Wait. What? You yell "Applebee's" at the GPS. You plead "Applebee's." But you are ignored. Your system responds calmly, telling you instead about the all-you-can-eat breadsticks at Olive Garden.

Perhaps that is a rather trivial scenario. Some might insist that corporate Olive Garden versus corporate Applebee's isn't really a choice anyway. But you get the idea. Control at the point of network access is critically important.

The critical technical elements of EI, the highly flexible mobile interface and the computational sophistication, are a decade or so in the future. Say what you will. These developments are going to happen. I've labeled them more or less inevitable. But the political, social, and economic expectations about who will control the interface will be established now—as soon as we recognize the interface will exist and how important it will be. The issue of control is up for grabs. There is work to be done.

Corporate Interface Capture

Back in the late 1990s and early 2000s, you couldn't open a magazine or your mailbox without an America Online floppy disk falling out. Slip the disk into your computer and connect up to this newfangled thing called the Internet through your phone for only $19.95 a month. As we noted earlier, at its peak, in the United States, AOL was the Internet interface for 35 million customers (just under half the total at that time), many of whom basically equated AOL with the Internet itself. The rest of the story is well known. In 2000, Time Warner and AOL merged in what turned out to be a multibillion-dollar disaster. AOL receded from dominance as dial-up connectivity was replaced by high-speed broadband provided primarily through the cable companies. Time Warner, a content company, saw the future of delivery as online and wanted to vertically integrate—to own the content and the network access point as well. It was a sound business strategy except that AOL turned out to be the wrong means to that end because it lost its access-near-monopoly to its broadband competitors.

Controlling both network access and network content was not just Time Warner's idea. AOL was already there. We have described its vision of a "walled garden." It is a provocative turn of phrase. The garden part refers to attractive content, and the walled part denotes that you are stuck there—no getting out. Think about it from AOL's point of view. Why only charge the retail customers for access to content when you could charge commercial firms for access to your millions of customers? For that to work, of course, you have to blockade the potential competition. During the 1990s, CBS paid to provide sports content, ABC paid to provide news, and 1–800-Flowers paid to be the default florist for anyone seeking one among AOL's customer base. This strategy became highly profitable for a time, but ultimately it failed when AOL lost its monopoly grip.

In the decades ahead, most of the likely players seeking dominance and control of the access technologies are already in the game —AT&T, Verizon, Comcast, Amazon, Facebook, Google, Microsoft, Apple, among others.

We see the controlled access principle successfully executed by Apple for access to its mobile operating system. *The New York Times* tells this story (one of many similar stories) from the summer of 2020:

> Executives and engineers from Facebook's games division submitted their new app, *Facebook Gaming,* to Apple last month for approval to offer it in the iPhone maker's App Store. Apple considered Facebook's application for a few weeks. This month, it delivered its verdict: denied. The Facebook team was not surprised. It wasn't the first time Apple had said no to the Facebook Gaming app. Or the second. Or even the third. Since February, Apple has rejected at least five versions of Facebook Gaming, according to three people with knowledge of the companies, who spoke on the condition of anonymity because the details are confidential. Each time, the people

> said, Apple cited its rules that prohibit apps with the "main purpose" of distributing casual games. Facebook Gaming may also have been hurt by appearing to compete with Apple's own sales of games, two of the people said. Games are by far the most lucrative category of mobile apps worldwide. Apple's App Store, the only officially approved place for iPhone and iPad users to find new games and other programs, generated about $15 billion in revenue last year.

Although it might strike you that controlling access to gaming apps like Angry Birds is relatively harmless, the underlying dynamic is anything but. The metric of success in the digital age is controlling access to public attention. The new billionaires of our age are not the newly crowned princes of timber, rails, oil, or even silicon chips. They are what Tim Wu calls the attention merchants—the likes of Mark Zuckerberg of Meta/Facebook, Larry Page and Sergey Brin of Google, Jack Dorsey (now Elon Musk) of Twitter, and Reed Hastings of Netflix. These men control platforms that steer our attention in an economic dynamic for which they are compensated spectacularly. When Steve Jobs at Apple and Jeff Bezos at Amazon saw the potential profitability of steering content through their systems, they started iTunes and Prime, respectively, to get in on the action. AT&T, Comcast, and Verizon owned platforms, saw the writing on the wall, and did the same. There are enough players and enough competition among the competing systems currently that the individual consumer seems to have a fair amount of choice and control, although getting access to what you want can get expensive.

Let's say Google becomes convinced the current political administration is going to save the day for some reason and secretly decides to skew search results and its YouTube recommendations and news feeds to convince you of the same. Google's dominance of online search is based on the continued conviction of its users that they are getting access to what they want, not what somebody else wants. There would be an uproar of protest and a very public scandal if search results were suddenly hijacked. There are dozens of viable search competitors. Google's multibillion-dollar viability depends on this continuing competitive dynamic, and its senior management is well aware of that. When Page and Brin decided to step back and put someone else in as CEO they picked Sundar Pichai. Why him? His long-standing job was managing the fine-tuning of the search algorithm to make sure users didn't become dissatisfied and jump ship—arguably the most important job at Google. So Google would sneak self-serving filters into its search algorithms at great peril to its franchise.

So far so good. Vibrant competition for audience attention. It is not yet the Orwellian thought police. But let's step back again and review what happens when you add the intelligent technologies of EI to the mix.

Three issues merit consideration here. First, the stakes are higher. EI has the potential to influence the character of our life decisions, careers, investments, and personal relationships. This is more significant than the current dynamics of which news feed or movie channel you have chosen. The incentives for commercial players to bias the system are very high. Second, the embedded biases may be harder to identify and to correct. Computational recommendations may be based on literally thousands of considerations, each influence-weighted according to the personal preferences and goal algorithms of the individual. Third, individuals may be lulled into knowingly accepting corporate biases. Making sure your evolutionary intelligence is fully reviewed and aligned will require time and effort. We may encounter something a little like present-day corporate loyalty "reward card" programs. Rather than think about it, just buy Acme brand products. They seem OK. They promise your price has been discounted, plus there are those reward points. I mention these rewards programs because in all likelihood they will morph in time to become increasingly electronic and sophisticated—that is, more EI-like.

Did you know that the U.S. standard railroad gauge measures 4 feet 8.5 inches? This is the width between the rails. It would appear to be an odd choice. Railroad aficionados will proudly explain that that it was the military specification for chariots in ancient Rome. The heavily rutted roads of the Roman Empire required standardized chariots. How did that military spec survive two millennia? The answer in part is that the first rail lines in the United Kingdom were built by the same people who built the pre-railroad tramways, which used that gauge. It turns out the people who built the tramways used the same jigs and tools for building wagons, which also used that traditional wheel spacing. Funny thing about how these traditions persist. My argument in this chapter is that taken-for-granted cultural traditions that we have come to accept today will likely get baked into the way EI gets defined in the decades ahead. Beware rutted roads.

My policy recommendation here, as it has been throughout, is for an explicit principle of transparency. If EI processes involving thousands of input variables become equivalent to the iconic black box, then use independent computational intelligence to test the reliability and honesty of the black box's outputs. If you think your Acme affiliation is inflating your costs or biasing your decisions, test it against another system. It is likely that as you are reading this book some security software you have installed is whirring away on your laptop and cell phone in the background to search for intruders and malware. If there be dragons with black hats in our future, there will be knights with white hats as well. Rather than just hope for white-hatted knights, let's anticipate the need for them and establish that need in policy.

Political Interface Capture

We have come to use the word "Orwellian" to describe a future in which the government controls our every thought and desire. George Orwell's *1984*, written just after the Second World War, was an artful mix of not-so-subtle references to both Hitler's Nazi Germany and Stalin's Soviet Russia. In the decades following its publication, *1984* would sell tens of millions of copies and be translated into more than sixty different languages, the greatest number for any novel at the time. We continue to use the terms "Big Brother" and "thought police" in common parlance today to conjure up these enduring themes. Like the imagery of Dr. Frankenstein's monster, Orwell's bad dream resonated and continues to resonate in public consciousness. Is it possible that evolutionary intelligence will come to be government-issue software?

We have a present-day model for what that might look like in "Kwangmyong," which is Korean for "Bright." It is North Korea's Orwellian answer to what would otherwise be the Internet. It is a self-contained, tightly controlled intranet, a tiny microcosm of the web with about 5,000 websites available only in North Korea. North Koreans have no access to the web outside. Exchanges on this network are carefully monitored, but the monitoring is seldom necessary. The traditional mass media in North Korea, of course, are equally tightly controlled, and most North Koreans appear to accept the oft-repeated message that the outside world and its ideas are dangerous and malevolent. There are few other examples of such total control. Even Cuba has relaxed its control over Internet access and retreated from its conviction that exposure to the outside world would lead to instant insurrection. Most North Koreans have never experienced anything like an open democratic culture, so the calm acceptance of Kwangmyong may be a special case. Is it possible that such state control over communication and commerce could be achieved gradually and perhaps surreptitiously?

Here the most worrisome case study may be the evolving Chinese social credit system. Still in the development stage, it combines familiar Western-style credit scores with online payment records and evaluations from neighbors and local police. Those on the approved "red list" receive perks such as discounts on heating bills, non-deposit bike rentals, and favorable bank loans. Those on the "black list" are refused train and plane tickets and hotel reservations and may have their pictures posted online or at movie theaters or libraries for an exercise in very public shaming. What gets you blacklisted? Certainly, official ties with the heretical Falun Gong movement are not a good idea. Other offenses leading to blacklisting are jaywalking, driving under the influence, excessive video gaming, criticizing the government, making late payments, failing to sweep the sidewalk in front of your house, or playing loud music on public transport.

It is not yet a national system backed up by facial recognition and artificial intelligence software, but that appears to be part of the plan. In 2014, the Chinese State Council's founding document explained that the idea was to "allow the trustworthy to roam everywhere under heaven while making it hard for the discredited to take a single step." Maoist abuses especially during the Cultural Revolution led to widespread distrust and cynicism among Chinese citizens. This new system is supposed to make mutual trust possible again. A surprisingly large proportion of the Chinese population including the well-educated and well-to-do are supportive of these systems—there is 80 percent overall public approval, as determined by one recent study. You can work your way off the blacklist by paying bills on time and doing community service. It's designed to motivate "good behavior," and so far, it is generally judged to be successful at that goal by authorities.

Some China watchers believe that horrified Western journalists are overreacting at what they see as Big Brother in action as these social credit systems evolve. Of course, there is public approval of such systems, they argue. This is simply what you get when you mix Chinese culture with big data. These problematic prospects may not be limited to China. Similar although smaller-scale models have been proposed in Russia, Germany, and the United Kingdom.

Transparency

We conclude with a return to one of our central themes. Technology is not in itself a force for good or a force for evil, nor is it neutral. Evolutionary intelligence is human intelligence, technically augmented. It can amplify humanity's highest ideals. It can moderate some of humanity's most dramatic weaknesses. But will it?

My policy proposal for increasing the odds that these technologies will move in the direction of these high ideals is straightforward—use these advanced technologies to monitor themselves. This is what I mean by transparency. If powerful corporate entities or governments are self-interestedly biasing the flow of information and the data on which important decisions are made, make those biases evident and public.

Currently the notion of computational transparency is associated with the idea of open-source software: Source code becomes publicly available so analysts can see what's going on. That's a good start. But as these platforms grow to millions of lines of code, the practical accessibility of transparency recedes. The key to testing computational decision algorithms is to test them—to run them through their paces and systematically compare them with competitors. Think of it as Consumer Reports for software systems. Underwriters Laboratories. ISO 9000 standards. *PC Magazine*'s Top Ten. Public audits. The more institutions in the transparency business the better.

The careful study of the past is our best guide to the future. Gordon Moore, as noted previously, observed back in the mid-1960s that the number of transistors on a microchip doubled about every two years as the corresponding cost of equivalent computers was halved. It seemed reasonable to project that observation into the future, and what became "Moore's law" proved to be a stunningly accurate prediction of the scale of technical evolution. Computers moved from immense walls of technology in large air-conditioned rooms to tiny chips in our phones and earpieces. And those tiny chips had exponentially more computational power than their predecessors. And, importantly, those tiny chips gradually became connected to each other in a single worldwide network. Moore's law predicted the powerful chips but not the powerful connectivity.

Currently, all those chips are embedded in our environment and are able to respond if we approach them. They are a few feet away in smart speakers or electronic kiosks. They are in our lap as a laptop or in our hand as a smartphone. We can walk up to a kiosk or take out a smartphone and click, talk, and type to initiate communication with the network of trillions of chips that surround us. "Hey Siri, what is the weather forecast?" "ATM, I want $200 from my account." "Facebook friends, look at my silly cat." But these interfaces are primitive and require some effort. We have to decide to take out our cell phone. We have to approach the kiosk.

So far Moore's little chips have moved from computer rooms to the boxes on our desks and phones in our hand. Next they will move to our clothes, our earpieces, our glasses and contact lenses. No longer are they just out there; rather, they are part of our capacity to understand and interact with what is out there. They are an extension of our senses. An extension of our ability to reason and to calculate the relationship between available means and imagined ends.

Some of today's technologies already use a form of digital intelligence in telling ways, including self-configuration, universal access to information, reputational feedback, intelligent filtering, and instant machine translation. Why have these innovations become successful? They empower our capacities: Each may add value or update our calculations with the latest information. These technologies generally don't intrude but wait patiently to be called on. Many draw on collective experience. Importantly, they improve as they learn from experience. These capacities are converging and moving closer to us.

The jump of computational intelligence from the box on the desk to our fingertips and glasses may sound scary. Or it may sound like no big deal. But it is both. And, I promise, in a few decades we will simply take all of this for granted. I personally don't worry about the scary part. I do worry about the no-big-deal part.

The evolution and structural control of evolutionary intelligence is in our hands if we recognize its significance. Let's continue to think this through, to argue about alternative pathways, to use our new tools to correct our inevitable mistakes. Evolutionary intelligence will not appear ten years out in a lightning flash and a gust of wind. Evolutionary intelligence is already here.

Acknowledgments

I would like to thank Marko Skoric, Nat Poor, Roei Davidson, Victor Cassella, Lee Rainie, Klaus Bruhn Jensen, John Markoff, Ben Shneiderman, Susan Douglas, David Weinberger, Peter Bernstein, Elizabeth Kaplan, Alison MacKeen, Yoav Bergner, Jan Plass, Stuart Lacy, Paul DiMaggio, Gita Manaktala, Sandra Braman, and Andy Lippman for comments, criticism, and encouragement. Special thanks to Susan and Max for help and support on this journey.

Notes

Prologue

xi The paradox of the brilliant inventor—For examples of inventors with an unsure grasp of how their inventions would actually be used, see Andrew Pettegree, *The Book in the Renaissance* (New Haven, CT: Yale University Press, 2010); Sidney H. Aronson, "Bell's Electronic Toy: What's the Use? The Sociology of Early Telephone Usage," in *The Social Impact of the Telephone*, ed. Ithiel de Sola Pool (Cambridge, MA: MIT Press, 1977), 15–39; Erik Barnouw, *A Tower in Babel* (New York: Oxford University Press, 1966); "Thomas J. Watson," Wikipedia; Janet Abbate, *Inventing the Internet* (Cambridge, MA: MIT Press, 1999).

xiii It is a rare moment in cultural history—Daniel Bell's full commentary on the unrealized significance of the Industrial Revolution can be found in his introduction to Simon Nora and Alain Minc, *The Computerization of Society* (Cambridge, MA: MIT Press, 1980).

xiii There is no reason for any individual to have a computer in his home—For Ken Olsen's famous quotation in context, see the Quote Investigator: http://quoteinvestigator.com/2017/09/14/home-computer/.

Chapter 1

8 Do drivers pay attention—You may have pondered the 99.48 percent figure. The source, for what it's worth, is "Motorist Compliance with Standard Traffic Control Devices," FHWA-RD-89–103, U.S. Department of Commerce National Technical Information Service, April 1989, https://rosap.ntl.bts.gov/view/dot/25685/dot_25685_DS1.pdf?.

8 Polish mounted cavalry—The Polish cavalry charge at Krojanty actually happened but was much more complicated than the myth. The 18th Pomeranian Uhlan Cavalry Regiment successfully attacked a German infantry battalion, delaying the German advance. The cavalry suffered heavy losses and retreated when German armored cars arrived with heavy machine guns. The photographs of fallen soldiers and horses led to the myth.

10 Enter Daniel Kahneman—For more background on prospect theory, see Amos Tversky and Daniel Kahneman, "Judgment under Uncertainty: Heuristics and Biases," *Science* 185, no. 4157 (1974): 1124–1131.

11 A global village—More on Marshall McLuhan's notion can be found in Maria Popova, "Uncovered Gem: Marshall McLuhan's Global Village," *The Marginalian* (formerly Brain Pickings), March 15, 2010, http://brainpickings.org/2010/03/15/marshall-mcluhan-global-village/#:~:text=The%20world%20is%20now%20like,away%20go%20the%20drums%20again.

11 In *Walden,* Thoreau grumbled—This particular passage appears on page 36 of the Houghton Mifflin 1976 edition.

13 On the very last page—W. Ross Ashby, *An Introduction to Cybernetics* (London: Chapman & Hall (1970), 272.

14 Douglas Engelbart (the inventor of the computer mouse)—Engelbart is a hero to many of us who see computational intelligence as complementary rather than necessarily in opposition to human agency. The quotation is from his writing on "Augmenting Human Intellect: A Conceptual Framework," Summary Report AFOSR-3223, Stanford Research Institute, 1962.

17 Psychologist Gary Marcus notes—His commentary on the natural ambiguity of human speech is from "Am I Human? Researchers Need New Ways to Distinguish Artificial Intelligence from the Natural Kind," *Scientific American* 316, no. 3 (March 2017).

17 Potential biases in facial recognition algorithms—See William Crumpler, "The Problem of Bias in Facial Recognition," Center for Strategic & International Studies, May 1, 2020, http://csis.org/blogs/technology-policy-blog/problem-bias-facial-recognition.

17 Federally sponsored blue ribbon committee—The earnest but unfortunate Pierce committee report (formally named the Automatic Language Processing Advisory Committee, 1966) can be found at http://pangeanic.com/knowledge_center/machine-translation-alpac-report/.

19 Roughly a third of transactions—The figures on cash and credit/debit card transactions are from G4S Cash Solutions, *World Cash Report 2018* (Utrecht: The

Netherlands, 2018), http://cashessentials.org/app/uploads/2018/07/2018-world-cash-report.pdf.

21 At the MIT Media Lab, for example, Pattie Maes and Rosalind Picard—See Pattie Maes and Pranav Mistry, "Meet the Sixthsense Interaction," TED Talk, 2017, and Pattie Maes, "Embracing Our Cyborg Selves," presentation at the MIT AR in Action Conference, February 8, 2017. For Rosalind Picard's perspective, see her book *Affective Computing* (Cambridge, MA: MIT Press, 1997).

22 Inventor and author Ray Kurzweil—The classic statement is from Ray Kurzweil, *The Singularity Is Near: When Humans Transcend Biology* (New York: Penguin Books, 2006) page 14. A new revised and expanded version is in the works, due out shortly.

Chapter 2

33 Mojo Vision—See Mark Sullivan, "The Making of Mojo, AR Contact Lenses That Give Your Eyes Superpowers," *Fast Company*, January 16, 2020, http://fastcompany.com/90441928/the-making-of-mojo-ar-contact-lenses-that-give-your-eyes-superpowers.

37 The tactile sensory path—Tactile communication is a neglected research area. For further details, see Ben Challis, "Tactile Interaction," chapter 20 in *The Encyclopedia of Human-Computer Interaction*, 2nd ed., International Design Foundation, 2014, http://interaction-design.org/literature/book/the-encyclopedia-of-human-computer-interaction-2nd-ed/tactile-interaction.

39 Cosmologist Carl Sagan famously remarked—His famous comment on the complexity of the human brain is discussed in Robert A. Snyder, *The Social Cognitive Neuroscience of Leading Organizational Change* (New York: Routledge, 2016).

Chapter 3

46 Intelligence is "not merely book learning"—This widely noted perspective on intelligence comes from "Mainstream Science on Intelligence," a public statement by a group of psychologists in *The Wall Street Journal* on December 13, 1994, in response to the controversy over *The Bell Curve* by Richard Herrnstein and Charles Murray.

49 Demonstrable systematic distortions in human cognition—The number of identifiable examples of "hardwired" bias and patterns of self-deception in the evolved human cognitive system is intimidatingly large. One way of summarizing the pattern broadly is to examine the "fundamental attribution error." For more detailed examinations, see Doron Kliger and Andrey Kudryavtsev, "The Availability Heuristic and Investors' Reaction to Company-Specific Events," *Journal of Behavioral*

Finance 11, no. 1 (2010): 50–65; Justin Kruger and David Dunning, "Unskilled and Unaware of It: How Difficulties in Recognizing One's Own Incompetence Lead to Inflated Self-Assessments," *Journal of Personality and Social Psychology* 77, no. 6 (1999): 1121–1134; Solomon E. Asch, *Group Forces in the Modification and Distortion of Images* (Englewood Cliffs, NJ: Prentice-Hall, 1952); Charles G. Lord, Lee Ross, and Mark R. Lepper, "Biased Assimilation and Attitude Polarization: The Effect of Theories on Subsequently Considered Evidence," *Journal of Personality and Social Psychology* 37, no. 11 (1979): 2098–2109; and George A. Miller, "The Magical Number Seven, Plus or Minus Two: Some Limits on Our Capacity for Processing Information," *Psychology Review* 63 (1956): 81–97.

59 As Fiske and Taylor explain—Susan T. Fiske and Shelley E. Taylor, *Social Cognition*, 3rd ed. (Thousand Oaks, CA: SAGE Publications, 2017), 59.

62 Uninformed thinkers are problematic—For further examples of widespread ignorance and misperception, see Ilya Somin, *Democracy and Political Ignorance: Why Smaller Government Is Smarter* (Stanford, CA: Stanford University Press, 2013). For an analysis from the committee of National Academies of Sciences, Engineering, and Medicine, see *Science Literacy: Concepts, Contexts, and Consequences* (Washington, DC: The National Academies Press, 2016), https://doi.org/10.17226/23595. See also A. S. Levy, S. B. Fein, and M. Stephenson, "Nutrition Knowledge Levels about Dietary Fats and Cholesterol: 1983–1988," *Journal of Nutrition Education* 25, no. 2 (1993): 60–66, and from the U.K., the "Health Education Monitoring Survey: Health of the Nation White Paper," Department of Health, 1992. For information on U.S. health literacy, see http://proliteracy.org. On financial literacy, see Justin McCarthy and Anita Pugliese, "Two in Three Adults Worldwide Are Financially Illiterate," Gallup, November18, 2015, https://news.gallup.com/poll/186680/two-three-adults-worldwide-financially-illiterate.aspx, and "Study: Americans Overestimate Their Financial Literacy," June 15, 2018, http://pymnts.com/consumer-finance/2018/study-americans-financial-literacy-personal-finance-knowledge/.

66 Confederate in a gorilla suit casually walks in among the players—For a discussion of the gorilla illusion, see Cathy N. Davidson, *Now You See It: How Technology and Brain Science Will Transform Schools and Business for the 21s t Century* (New York: Penguin Books, 2012), 301; and Roger Highfield, "Did You See the Gorilla?" Telegraph.co.uk, May 5, 2004. For more on the one percent of visual stimuli phenomenon, see Li Zhaoping, *Understanding Vision: Theory, Models, and Data* (Oxford: Oxford University Press, 2014).

67 Lippmann proceeds to describe the analysis of their "eyewitness" reports—Walter Lippmann's discussion of the German study is on page 55 of the Free Press edition of his classic book *Public Opinion*.

69 Participants read a description—The example is drawn from Linda J. Skitka, Elizabeth Mullen, Thomas Griffin, Susan Hutchinson, and Brian Chamberlin,

"Dispositions, Scripts, or Motivated Correction? Understanding Ideological Differences in Explanations for Social Problems," *Journal of Personality and Social Psychology* 83, no. 2 (2002): 473.

70 As Jost and Krochik summarize—John T. Jost and Margarita Krochik, "Ideological Differences in Epistemic Motivation: Implications for Attitude Structure, Depth of Information Processing, Susceptibility to Persuasion, and Stereotyping," *Advances in Motivation Science* 1 (2014): 181.

71 The classic game of this tradition is the prisoner's dilemma—See Elvis Picardo, "The Prisoner's Dilemma in Business and the Economy," *Investopedia*, January 22, 2020; and Charles B. Parselle, "No Way Out: Negotiation and the Prisoner's Dilemma," *Mediate.com*, April 2007.

74 This mindfulness model—See Scott R. Bishop et al., "Mindfulness: A Proposed Operational Definition," *Clinical Psychology: Science and Practice* 11, no. 3 (2004): 230–241.

77 In ancient Greece juries included up to 500 jurors—The fascinating notion of 500-person juries was drawn from Loren J. Samons, *The Cambridge Companion to the Age of Pericles* (Cambridge: Cambridge University Press, 2007), 244, 246.

81 Swiss psychologist Klaus Scherer—Scherer's oft-cited definition of emotion can be found in Klaus R. Scherer, "Unconscious Process in Emotion: The Bulk of the Iceberg," in *Emotion and Consciousness*, ed. L. F. Barrett, P. M. Niedenthall, and P. Winkielman (New York: Guilford, 2005), 314.

81 Tell that to Rosalind Picard—See Rosalind W. Picard, *Affective Computing* (Cambridge, MA: MIT Press, 1997), and Christina Couch, "Can AI Learn to Understand Emotions?" *NOVA on PBS*, May 16, 2018, http://www.pbs.org/wgbh/nova/article/affective-computing.

Chapter 4

91 A particularly breathless critic—Ellen Wartella and Byron Reeves, "Historical Trends in Research on Children and the Media: 1900–1960," *Journal of Communication* 35, no. 2 (1985): 118–133.

94 And among the most troubling of very sophisticated unreality is the deep fake—For more on this fast-developing domain, see http://regmedia.co.uk/2019/10/08/deepfake_report.pdf/. See also Robert M. Chesney and Danielle Keats Citron, "Deep Fakes: A Looming Challenge for Privacy, Democracy, and National Security," *California Law Review* 107 (2018): 1753. Ian Bogost, however, in the May 29, 2019, issue of *The Atlantic*, notes in "Facebook's Dystopian Definition of 'Fake'" that "for the social-media platform, a doctored video of Nancy Pelosi is content, not a phony."

95 Robert Putnam captured public attention—Robert D. Putnam, *Bowling Alone: The Collapse and Revival of American Community* (New York: Simon & Schuster, 2000).

96 Nicholas Negroponte's now-famous technological notion of the "Daily Me"—Nicholas Negroponte, *Being Digital* (New York: Knopf, 1995).

97 The Israeli historian and intellectual celebrity Yuval Noah Harari—Yuval N. Harari, *21 Lessons for the 21st Century* (New York: Spiegel & Grau, 2018), 81.

98 Google's annual advertising revenues—Another moving target. This estimate is drawn from "Google: Global Annual Revenue 2002–2021," *Statista, com*, last updated December 2, 2022, https://www.statista.com/statistics/266206/googles-annual-global-revenue/. By 2018, Google's revenues were over $130 billion and increased to over $250 billion in 2021.

98 Routinely monetized their likenesses—It turns out that if you are "famous," you can cash in on the fame itself. See Cody Reaves, "Show Me the Money: Determining a Celebrity's Fair Market Value in a Right of Publicity Action," *Michigan Journal of Law Reform* 50, no. 831 (2017).

99 The typical marketing cost for major purchases—The sources for the statistics on car marketing and perfume, respectively, are Charles Morris, "Auto Industry (except Tesla) Spends an Average $1,000 per Vehicle in Advertising," *Charged Electric Vehicles Magazine*, July 15, 2016, https://chargedevs.com/newswire/auto-industry-except-tesla-spends-an-average-1000-per-vehicle-in-advertising/#:~:text=According%20to%20Chowdhry%2C%20the%20industry,in%20advertising%20per%20vehicle%20sold, and Paddy Calistro, "$150 for $1.50 Worth of Perfume," *LA Times*, November 13, 1988, http://latimes.com/archives/la-xpm-1988-11-13-tm-10-story.html.

101 Privacy-oriented activists—The ongoing battles between privacy advocates and online marketers is well documented in Lily Hay Newman, "The Fractured Future of Browser Privacy," *Wired*, January 30, 2020; and Kashmir Hill, "'Do Not Track,' the Privacy Tool Used by Millions of People, Doesn't Do Anything," *Gizmodo*, October 15, 2018.

102 In 1840, roughly one in ten Americans lived in urban areas—The urbanization trend data is drawn from "Urbanization United States Summary: 2010 Census of Population and Housing, Population and Housing Unit Counts," U.S. Census Bureau, Washington, DC (2012).

106 Mobile phones have diffused—The review of mobile phone usage is drawn from Will Marler, "Mobile Phones and Inequality: Findings, Trends, and Future Directions," *New Media & Society* 20, no. 9 (2018): 3498. See also Yong Jin Park, "My Whole World's in My Palm! The Second-Level Divide of Teenagers' Mobile Use and Skill," *New Media & Society* 17, no. 6 (2014): 977–995.

107 Sociologist Eszter Hargittai—An overview of Hargittai's research on the second-order digital divide can be found in Eszter Hargittai and Yu-li Patrick Hsieh, "Digital Inequality," in *Oxford Handbook of Internet Studies*, ed. W. H. Dutton (New York: Oxford University Press, 2013), 129–150.

108 Erik Brynjolfsson and colleagues point out—Erik Brynjolfsson, Tom Mitchell, and Daniel Rock, "What Can Machines Learn, and What Does It Mean for Occupations and the Economy?" *AEA Papers and Proceedings* 108 (2018): 43–47.

109 More than 500 editions of the novel—The data on the cultural influence of the Frankenstein meme is drawn from Susan Tyler Hitchcock, *Frankenstein: A Cultural History* (New York: W. W. Norton, 2007).

110 Alan Turing said the following about independent machine intelligence—Alan M. Turing, "Lecture to the London Mathematical Society," February 20, 1947.

110 Turing's friend and colleague—Irving J. Good, "Speculations concerning the First Ultraintelligent Machine," *Advances in Computers* 6 (1965).

110 British philosopher Nick Bostrom expanded on his oft-repeated paper clip scenario—See Kathleen Miles, "Artificial Intelligence May Doom the Human Race within a Century," *Huffington Post*, February 4, 2015. See also Nick Bostrom, "Existential Risks: Analyzing Human Extinction Scenarios and Related Hazards," *Journal of Evolution and Technology* 9, no. 1 (2002).

111 Stephen Hawking, writing with colleagues—Stephen Hawking, Stuart Russell, Max Tegmark, and Frank Wilczek, "Transcendence Looks at the Implications of Artificial Intelligence—but Are We Taking AI Seriously Enough?" *Independent*, May 1, 2014.

112 Both the administrations of Barack Obama and Donald Trump—Barack Obama, Joi Ito, and Scott Dadich, "Barack Obama, Neural Nets, Self-Driving Cars, and the Future of the World," *Wired*, October 2016.

114 Private and public investments in AI—AI investment data is drawn from Paul Sawers, "Tech Nation: U.S. companies Raised 56% of Global AI Investment since 2015, Followed by China and the U.K.," *VentureBeat*, March 16, 2020, http://venturebeat.com/2020/03/16/tech-nation-u-s-companies-raised-56-of-global-ai-investment-since-2015-followed-by-china-and-u-k/, and Bergur Thormundsson, "Artificial Intelligence Funding United States 2011–2019," *Statista.com*, March 17, 2022, http://statista.com/statistics/672712/ai-funding-united-states/. This is another one of those moving-target estimates likely to be out of date within minutes of release.

116 A classic case is COMPAS software—The information on the COMPAS case is drawn from Julia Angwin, Jeff Larson, Surya Mattu, and Lauren Kirchner, "Machine Bias," *ProPublica*, May 23, 2016, http://propublica.org/article/machine-bias-risk-assessments-in-criminal-sentencing. See also Natasha Singer and Cade Metz, "Many

Facial-Recognition Systems Are Biased, Says U.S. Study," *New York Times*, December 19, 2019.

117 Google and its search subsidiaries—The current estimate of Google's dominance of search is drawn from "Global Market Share of Search Engines 2010–2022," *Statista.com*, last updated December 1, 2022, http://statista.com/statistics/216573/worldwide-market-share-of-search-engines/.

Chapter 5

121 Modern speech is only about 100,000 years old—The estimates on the evolution of human speech is drawn from Robert C. Berwick and Noam Chomsky, *Why Only Us: Language and Evolution* (Cambridge, MA: MIT Press, 2016).

122 Settled on land for only 10,000 years—There is predictable scholarly dispute on this point, but most historians agree with the 10,000 year estimate. For an exemplar of this literature, see Allen W. Johnson and Timothy K. Earle, *The Evolution of Human Societies: From Foraging Group to Agrarian State* (Stanford, CA: Stanford University Press, 2000).

122 The printing press of the fifteenth century—The mass literacy estimates are drawn from James Westfall Thompson, *The Literacy of the Laity in the Middle Ages* (Berkeley, CA: University of California Press, 1939).

124 Tomasello puts it this way—Michael Tomasello, *The Cultural Origins of Human Cognition* (Cambridge, MA: Harvard University Press, 1999), 2–4.

125 Life as a hunter-gatherer was largely a desperate nomadic search for food—The classic characterization of hunter-gatherers is drawn from Graeme Barker, *The Agricultural Revolution in Prehistory: Why Did Foragers Become Farmers?* (Oxford: Oxford University Press, 2009). The standard view has been challenged in David Graeber and David Wengrow, *The Dawn of Everything: A New History of Humanity* (New York: Farrar, Straus and Giroux, 2021). They argue there is evidence that human life in this era was much more creative, collaborative, and less "desperate" than believed in the classic view.

127 An observation about the Neolithic revolution—Jared Diamond, *Guns, Germs, and Steel: The Fates of Human Societies* (New York: Norton, 1997), 106.

128 Stark social dynamics—Peter J. Richerson, Robert Boyd, and Robert L. Bettinger, "Was Agriculture Impossible during the Pleistocene but Mandatory during the Holocene? A Climate Change Hypothesis," *American Antiquity* 66, no. 3 (2001): 387–411.

129 Report breathlessly—This characterization of the Industrial Revolution is an amalgamation of several Wikipedia articles.

131 Gutenberg and his press are easy to pinpoint—The much-lauded review of this history is Elizabeth L. Eisenstein, *The Printing Press as an Agent of Change* (Cambridge: Cambridge University Press, 1979). See also Lucien Febvre and Henri-Jean Martin, *The Coming of the Book: The Impact of Printing, 1450–1800* (London: Verso, 2010).

133 Reading and writing for the mass population—Carl F. Kaestle, "The History of Literacy and the History of Readers," *Review of Research in Education* 12, no. 1 (1985): 11–53.

134 The growth of the middle class—Jon Miltimore, "The Growth of the World's Middle Class May Be the Greatest Story of Our Age," Foundation for Economic Education, August 29, 2018; and Homi Kharas, "The Unprecedented Expansion of the Global Middle Class," Brookings Global Economy and Development Working Paper 100, February 2017.

134 The significance of the revolution of the middle class—Max Weber, *Economy and Society* (Berkeley: University of California Press, 1978). First published in 1924.

Chapter 6

141 A December day in Silicon Valley—The story of Steve Jobs at Xerox PARC has become legend. See Malcolm Gladwell, "Creation Myth," *The New Yorker*, May 9, 2011; see also Walter Isaacson, *Steve Jobs* (New York: Simon & Schuster, 2011).

148 Clever tricks of car salesmen—I'm not a car guy, so I make a point of avoiding car shows and auto sales rooms. Not so for James Bragg. See his book *Letting the Cat Out of the Bag* (self-pub., CreateSpace, 2014), and his *Fighting Chance* website (https://fightingchance.com/).

152 On Labor Day Weekend—Also of mythic status is the broken laser pointer. See Adam Cohen, *The Perfect Store: Inside eBay* (New York: Back Bay Books, 2008).

156 Fraud on eBay—This estimate comes from Ronald J. Bauerly, "Online Auction Fraud and eBay," *Marketing Management Journal* 19, no. 1 (2009): 134–144.

158 Economic incentives remain strong—Brian McWilliams, *Spam Kings: The Real Story Behind the High-Rolling Hucksters Pushing Porn Pills and @#?% Enlargements* (Sebastopol, CA: O'Reilly Media, 2014).

161 This amazing technical achievement—Wikipedia has an excellent overview of "Machine Translation,", https://en.wikipedia.org/wiki/Machine_translation.

163 AT&T, which called its invention ISDN—Erika Andersen, "It Seemed Like a Good Idea at the Time," *Forbes*, October 4, 2013.

165 Largely an accident—For more on the happy accident that became the Internet, see Janet Abbate, *Inventing the Internet* (Cambridge, MA: MIT Press, 1999).

173 Complex algorithms like Google search—See Brian Dean, "Google's 200 Ranking Factors: The Complete List (2022)," *Backlinko*, updated October 10, 2021, http://backlinko.com/google-ranking-factors.

Chapter 7

179 Medical decision making—The review of evolving medical technology draws on an article by Elizabeth Gardner, "DNA Will Point the Way to Healthier Patients," *U.S. News and World Report*, November 8, 2017, http://usnews.com/news/healthcare-of-tomorrow/articles/2017-11-08/dna-will-point-the-way-to-healthier-patients.

180 Our behavior has a major influence on our health—The smoking data is from the American Lung Association, "Tobacco Trends Brief," accessed December 2, 2022, http://lung.org/research/trends-in-lung-disease/tobacco-trends-brief.

181 Sociologist Mark Granovetter's famous insight—Granovetter's article "The Strength of Weak Ties," *American Journal of Sociology* 78, no. 6 (1973): 1360–1380 is a true classic.

182 Don't people lie—Jeffrey Hancock, Catalina Toma, and Nicole Ellison, "The Truth about Lying in Online Dating Profiles," presentation at the Conference on Human Factors in Computing Systems, CHI 2007, San Jose, CA, April 28–May 3, 2007.

183 Web-mediated romance—The dating data is from Michael Rosenfeld, Sonia Hausen, and Reuben J. Thomas, "Disintermediating Your Friends: How Online Dating in the United States Displaces Other Ways of Meeting," *Proceedings of the National Academy of Sciences* 116, no. 36 (2019).

183 One observer estimated—The estimate that 80 percent of legal work is standardized and highly routine comes from Erik P. M. Vermeulen, "The Rise of Lawyers in a Digital World," *HackerNoon*, January 11, 2019, http://hackernoon.com/the-rise-of-lawyers-in-a-digital-world-a3098353c24c.

189 But as Daniel Susskind explains—Daniel Susskind, *A World without Work: Technology, Automation, and How We Should Respond* (New York: Metropolitan Books/Henry Holt, 2020).

Chapter 8

193 A convergence of biology and technology—The classic view here is Ray Kurzweil, *The Singularity Is Near: When Humans Transcend Biology* (New York: Penguin Books, 2006). A new version is in the works.

195 Easily manufactured—Although manufacturing microchips is indeed printing on silicon and relatively inexpensive per unit at scale, our brief description is an

oversimplification that ignores the extraordinary expense of the design process, the fabrication plants, and the rarefied chemistries of the production process.

199 If you control network access—With my able colleagues, I have reviewed some stories of historical network economics in W. Russell Neuman, Lee McKnight, and Richard Jay Solomon, *The Gordian Knot: Political Gridlock on the Information Highway* (Cambridge, MA: MIT Press, 1998).

199 One observer called it—On the "license to print money" notion, see the story at the Television History Site, http://www.teletronic.co.uk/pages/history_of_itv_6.html#:~:text=It%20was%20Roy%20Herbert%20Thomson,named%20Scottish%20Television)%20in%201957.

200 An America Online floppy disk—For more on the intriguing AOL case history, see Dan Tynan, "The 25 Worst Tech Products of All Time," *PCWorld*, May 26, 2006. The estimate of AOL's dominant share comes from Susan P. Crawford, *Captive Audience: The Telecom Industry and Monopoly Power in the New Gilded Age* (New Haven, CT: Yale University Press, 2012). See also Tim Wu, *The Attention Merchants: The Epic Scramble to Get Inside Our Heads* (New York: Penguin Random House, 2016), 210.

201 The *New York Times* tells this story—Seth Schiesel, "Apple Rejects Facebook's Gaming App, for at Least the Fifth Time," *New York Times*, June 18, 2020, http://nytimes.com/2020/06/18/technology/apple-ios-facebook-gaming-app.html.

205 *1984* would sell tens of millions of copies—John Rodden, *The Politics of Literary Reputation: The Making and Claiming of "St. George" Orwell* (New York: Oxford University Press, 1989).

205 "Kwangmyong," which is Korean for "Bright"—Eric Talmadge, "North Korea: Where the Internet Has Just 5,500 Sites," *Toronto Star*, February 23, 2014, http://thestar.com/news/world/2014/02/23/north_korea_where_the_internet_has_just_5500_sites.html.

205 Chinese social credit system—Genia Kostka, "China's Social Credit Systems and Public Opinion: Explaining High Levels of Approval," *New Media & Society* 21, no. 7 (2019): 1565–1593.

Index